UCODE_2005 and Six Other Computer Codes for Universal Sensitivity Analysis, Calibration, and Uncertainty Evaluation

Constructed using the JUPITER API

JUPITER: Joint Universal Parameter IdenTification and Evaluation of Reliability
API: Application Programming Interface

By Eileen P. Poeter, Mary C. Hill, Edward R. Banta, Steffen Mehl, and Steen Christensen

Prepared in cooperation with the
U.S. Environmental Protection Agency and the
International Ground Water Modeling Center, Colorado School of Mines

Techniques and Methods 6-A11

U.S. Department of the Interior
U.S. Geological Survey

U.S. DEPARTMENT OF THE INTERIOR
Gale Norton, Secretary

U.S. GEOLOGICAL SURVEY
P. Patrick Leahy, Acting Director

U.S. Geological Survey, Reston, Virginia: 2005

For product and ordering information:
World Wide Web: http://www.usgs.gov/pubprod
Telephone: 1-999-ASK-USGS

For more information on the USGS – the Federal source for science about the Earth,
its natural and living resources, natural hazards, and the environment:
World WideWeb: http://www.usgs.gov/pubprod
Telephone: 1-888-ASK-USGS

Suggested citation:
Poeter, E.P., Hill, M.C., Banta, E.R., Mehl, Steffen, and Christensen, Steen, 2005, UCODE_2005
and Six Other Computer Codes for Universal Sensitivity Analysis, Calibration, and Uncertainty
Evaluation: U.S. Geological Survey Techniques and Methods 6-A11, 283p.

PREFACE

This report describes the capabilities and use of the computer code UCODE_2005 and six other computer codes.

UCODE_2005 and three of the other computer codes – RESIDUAL_ANALYSIS, LINEAR_UNCERTAINTY, and MODEL_LINEARITY – replace the computer code UCODE (Poeter and Hill, 1998). They can also be used in place of the MODFLOW-2000 Parameter-Estimation Process and part of the Observation Process, and the post-processing codes RESAN-2000, YCINT-2000, and BEALE-2000 (Hill and others, 2000).

UCODE_2005 used with the other three codes – RESIDUAL_ANALYSIS_ADV, MODEL_LINEARITY_ADV, and CORFAC_PLUS – provide nearly all the capabilities of the MODFLOW-2000 UNC Process described by Christensen and Cooley (2005).

MODFLOW-2005 does not include a Parameter-Estimation Process and the Observation Process is limited to the calculating simulated equivalents of observations. For MODFLOW-2005, methods for sensitivity analysis, calibration, and uncertainty evaluation can be obtained using the programs documented in this report, PEST (Doherty, 2004), OSTRICH (Matott, 2005), or other similar programs. The programs documented here are unique in some of their capabilities and in their being constructed using the JUPITER API. The latter is important in part because the input and output files are structured to encourage coordination with other programs constructed with the JUPITER API. See Appendix A for additional information.

The six other codes are not limited to use with UCODE_2005; the required input files can be created by any program with similar capabilities.

The documentation presented in this report includes brief listings of the methods used and detailed descriptions of the required input files and typical use of the output files. Detailed information on the methods, guidelines for conducting sensitivity analysis, data needs assessment, calibration, and uncertainty evaluation of a model of a complex system using examples mostly from ground-water modeling, and well-documented instructional exercises are presented by Hill and Tiedeman (in press). Together, this report, Hill and Tiedeman (in press), and Christensen and Cooley (2005) serve to document the computer codes described in this work. A more limited discussion of most of the methods and a previous version of the guidelines are presented in Hill (1994) and Hill (1998), which served to partially document the earlier programs. More on the methods described by Christensen and Cooley (2005) is presented by Cooley (2004).

This codes documented by the report are public domain, open-source software and can be downloaded from the Internet at URL http://water.usgs.gov/software/ground_water.html/.

The performance of the codes presented in this work has been tested in a variety of applications. Future applications, however, might reveal errors that were not detected in the test simulations. Users are requested to notify the originating office of any errors found in the report or the computer codes. Updates might occasionally be made to both the report and to the codes. Users can check for updates on the Internet at URL http://water.usgs.gov/software/ground_water.html/.

Contents

Figures

Tables

UCODE_2005 and Six Other Computer Codes for Universal Sensitivity Analysis, Calibration, and Uncertainty Evaluation

Constructed using the JUPITER API

By Eileen P. Poeter[1], Mary C. Hill[2], Edward R. Banta[3], Steffen Mehl[2], and Steen Christensen[4]

Abstract

This report documents the computer codes UCODE_2005 and six post-processors. Together the codes can be used with existing process models to perform sensitivity analysis, data needs assessment, calibration, prediction, and uncertainty analysis. Any process model or set of models can be used; the only requirements are that models have numerical (ASCII or text only) input and output files, that the numbers in these files have sufficient significant digits, that all required models can be run from a single batch file or script, and that simulated values are continuous functions of the parameter values. Process models can include pre-processors and post-processors as well as one or more models related to the processes of interest (physical, chemical, and so on), making UCODE_2005 extremely powerful. An estimated parameter can be a quantity that appears in the input files of the process model(s), or a quantity used in an equation that produces a value that appears in the input files. In the latter situation, the equation is user-defined.

UCODE_2005 can compare observations and simulated equivalents. The simulated equivalents can be any simulated value written in the process-model output files or can be calculated from simulated values with user-defined equations. The quantities can be model results, or dependent variables. For example, for ground-water models they can be heads, flows, concentrations, and so on. Prior, or direct, information on estimated parameters also can be considered. Statistics are calculated to quantify the comparison of observations and simulated equivalents, including a weighted least-

[1] International Ground Water Modeling Center and the Colorado School of Mines, Golden, Colorado, USA
[2] U.S. Geological Survey, Boulder, Colorado, USA
[3] U.S. Geological Survey, Lakewood, Colorado, USA
[4] Department of Earth Sciences, University of Aarhus, Aarhus, Denmark

squares objective function. In addition, data-exchange files are produced that facilitate graphical analysis.

UCODE_2005 can be used fruitfully in model calibration through its sensitivity analysis capabilities and its ability to estimate parameter values that result in the best possible fit to the observations. Parameters are estimated using nonlinear regression: a weighted least-squares objective function is minimized with respect to the parameter values using a modified Gauss-Newton method or a double-dogleg technique. Sensitivities needed for the method can be read from files produced by process models that can calculate sensitivities, such as MODFLOW-2000, or can be calculated by UCODE_2005 using a more general, but less accurate, forward- or central-difference perturbation technique. Problems resulting from inaccurate sensitivities and solutions related to the perturbation techniques are discussed in the report. Statistics are calculated and printed for use in (1) diagnosing inadequate data and identifying parameters that probably cannot be estimated; (2) evaluating estimated parameter values; and (3) evaluating how well the model represents the simulated processes.

Results from UCODE_2005 and codes RESIDUAL_ANALYSIS and RESIDUAL_ANALYSIS_ADV can be used to evaluate how accurately the model represents the processes it simulates. Results from LINEAR_UNCERTAINTY can be used to quantify the uncertainty of model simulated values if the model is sufficiently linear. Results from MODEL_LINEARITY and MODEL_LINEARITY_ADV can be used to evaluate model linearity and, thereby, the accuracy of the LINEAR_UNCERTAINTY results.

UCODE_2005 can also be used to calculate nonlinear confidence and predictions intervals, which quantify the uncertainty of model simulated values when the model is not linear. CORFAC_PLUS can be used to produce factors that allow intervals to account for model intrinsic nonlinearity and small-scale variations in system characteristics that are not explicitly accounted for in the model or the observation weighting.

The six post-processing programs are independent of UCODE_2005 and can use the results of other programs that produce the required data-exchange files.

UCODE_2005 and the other six codes are intended for use on any computer operating system. The programs consist of algorithms programmed in Fortran 90/95, which efficiently performs numerical calculations. The model runs required to obtain perturbation sensitivities can be performed using multiple processors. The programs are constructed in a modular fashion using JUPITER API conventions and modules. For example, the data-exchange files and input blocks are JUPITER API conventions and many of those used by UCODE_2005 are read or written by JUPITER API modules. UCODE-2005 includes capabilities likely to be required by many applications (programs) constructed using the JUPITER API, and can be used as a starting point for such programs.

Chapter 1: INTRODUCTION

Recent work has clearly demonstrated that inverse modeling and associated methods, though imperfect, provide capabilities that help modelers take greater advantage of the insight available from their models and data. Expanded use of this technology requires tools with different capabilities than those that exist in currently available inverse models. UCODE (Poeter and Hill, 1998) has two attributes that are not jointly available in other inverse models: (1) the ability to work with any mathematically based model or pre- or post-processor with ASCII or text-only input and output files, and (2) the inclusion of informative statistics with which to evaluate the importance of observations to parameters and the importance of parameters to predictions. To address the need to enhance inverse modeling and associated methods further, the U.S Geological Survey (USGS), in cooperation with the U.S. Environmental Protection Agency and the International Ground Water Modeling Center of the Colorado School of Mines, expanded the functionality of UCODE to produce the computer programs documented in this report:
UCODE_2005,
RESIDUAL_ANALYSIS,
RESIDUAL_ANALYSIS_ADV,
LINEAR_UNCERTAINTY,
MODEL_LINEARITY,
MODEL_LINEARITY_ADV, and
CORFAC_PLUS.

The programs presented in this report are designed to work with existing software packages (called process models in this work) that use numerical (ASCII or text only) input, produce numerical output, and can be executed in batch mode. Specifically, the programs were developed to do the following

(1) Manipulate process-model input files and read values from process-model output files.

(2) Compare user-provided observations with equivalent simulated values derived from the process-model output files using a number of summary statistics, including a weighted least-squares objective function.

(3) Use optimization methods to adjust the value of user-selected input parameters in an iterative procedure to minimize the value of the weighted least-squares objective function.

(4) Report the estimated parameter values.

(5) Calculate and print statistics used to (a) diagnose inadequate data or identify parameters that probably cannot be estimated, (b) evaluate estimated parameter values, (c) evaluate model fit to observations, and (d) evaluate how accurately the model represents the processes.

Process models executed by UCODE_2005 can include pre-processors and post-processors as well as models related to the processes of interest (physical, chemical, and so on), making UCODE_2005 extremely powerful. In general, graphical user interfaces cannot be used directly with UCODE_2005, but can be adapted with relatively little effort.

The programs documented here are constructed using conventions and modules of the JUPITER API (See Appendix A). The six other codes are not limited to use with UCODE_2005 – they can be used with any model that produces the required data-exchange files.

Purpose and Scope

This report documents how to use UCODE_2005, a universal inverse code, and the six other codes
RESIDUAL_ANALYSIS,
RESIDUAL_ANALYSIS_ADV,
LINEAR_UNCERTAINTY,
MODEL_LINEARITY,
MODEL_LINEARITY_ADV, and
CORFAC_PLUS.

These codes can be used with process models from any discipline, so readers of this report may come from many backgrounds. Different fields tend to have their own problems and literature related to inverse modeling. The reader is encouraged to become familiar with the literature in their field.

This report primarily documents the input, output, and execution of UCODE_2005 and the six other codes. A thorough description of the methods, including equations, guidelines for their use, example applications, and instructional exercises, are presented in other works, such as Hill and Tiedeman (in press), Christensen and Cooley (2005), Cooley (2004), and Cooley and Naff (1990).

This report begins with an overview of how UCODE_2005 solves nonlinear regression problems. The nonlinear regression methods used in UCODE_2005 and guidelines for their use in model calibration are described by Hill and Tiedeman (in press); a more limited description is available in Hill (1998). The regression theory is derived largely from Cooley and Naff (1990). Basic ideas from those works are presented briefly in this report. UCODE_2005 can use sensitivities produced by other programs or sensitivities it calculates using perturbation methods. Difficulties and solutions related to sensitivities are discussed.

Chapters 4 through 13 describe, in detail, how to run UCODE_2005 and construct input files. Chapter 14 describes the UCODE_2005 output files. UCODE_2005 produces data-exchange files that make results readily available for use by other codes. The data-exchange files are described in Chapter 14 and listed alphabetically with other program-produced files in Appendix B.

Input and output for the three codes RESIDUAL_ANALYSIS, LINEAR_UNCERTAINTY, and MODEL_LINEARITY are described in Chapter 15. Analysis of model fit and evaluation of uncertainty using linear methods is discussed in greater depth by Hill and Tiedeman (in press).

Chapter 16 discussed how to use the output files produced for checking simulated values, sensitivity analysis, parameter estimation, residual analysis, prediction and prediction uncertainty using linear methods, and for testing model linearity.

Chapter 17 describes the three codes RESIDUAL_ANALYSIS_ADV,
MODEL_LINEARITY_ADV, and CORFAC_PLUS. It also described how to use UCODE_2005 to calculate nonlinear confidence intervals.

Results are obtained by running UCODE_2005 and the other codes in appropriate sequences. The sequences are described using flowcharts and tables that show what files are produced and consumed at each step. For the six other codes documented in this work, most of the input files are generated by a preceding step; additional optional files can be provided by the user. The only other code that always needs a user-defined input file is CORFAC_PLUS. The files produced by the codes documented in this report are named using a filename prefix defined on command lines and filename extensions defined within the codes.

Appendix A describes the relation of UCODE_2005 and the other six codes to the JUPITER API. Appendix B lists the files produced using the filename prefix defined on the command line of each code. Appendix C includes selected input and output files from a process model, UCODE_2005, and other codes for a simple problem. Appendix D describes the directory structure of distributed files. Appendix E compares the capabilities of UCODE_2005 to those of UCODE (Poeter and Hill, 1998) to help UCODE users take advantage of the more recent version. Appendix F provides a condensed set of input instructions for UCODE_2005. Appendix G provides a condensed set of input instructions for the other codes with user-generated input files.

The expertise of the authors is in the simulation of ground-water systems, so examples in this report come from that field. The codes, however, have nearly unlimited applicability to problems with simulated values that are continuous functions of the parameter values.

Users of the codes presented in this work need to be familiar with the process model(s) and the simulated processes. In addition, although this report is written at an elementary level, some knowledge about basic statistics and the application of nonlinear regression is assumed. For example, it is assumed that the reader is familiar with the terms "standard deviation, variance, correlation, optimal parameter values, and residual analysis". Readers who are unfamiliar with these terms will understand this report better if they use a basic statistic book such as Helsel and Hirsch (2002) as a reference as they read this report. Useful references and applications are cited in Hill and Tiedeman (in press), including the illustrative example originally described by Poeter and Hill (1997). Hill (1998) provides a dated reference list.

Source files for UCODE_2005 and the post-processors are available at the Internet address http://water.usgs.gov/software/ucode.html/. The program distribution and installation are described in appendix D.

Acknowledgements

The authors would like to gratefully acknowledge Justin Babendreier of the U.S. Environmental Protection Agency for his good advice, wisdom, vision, and support for the first author; Richard Yager of the U.S. Geological Survey in Ithaca, New York for introducing the authors to the idea of a universal inverse code in the early 1990's; and John Doherty of Watermark computing and the University of Queensland whose contributions to the JUPITER API (Banta, Poeter, Doherty, and Hill, written commun., 2005) greatly influenced UCODE_2005. Parts of Chapter 11 are modified from Dr. Doherty's contribution to the JUPITER API documentation.

We would also like to acknowledge Michael LeFrancois, a student at the Colorado School of Mines in Golden, Colorado; Laura Foglia, a student at ETH in Zurich, Switzerland; and Charles Heywood and Claire Tiedeman of the U.S. Geological Survey in Grand Junction, Colorado and Menlo Park California, respectively, for providing ideas and testing the programs using their data sets. Claire Tiedeman's considerable effort deserves special note. Beta testing programs is always frustrating and these colleagues brought many good ideas and good humor to the process.

Chapter 2: OVERVIEW AND PROGRAM CONTROL

This section presents an overview of commonly used aspects of UCODE_2005, RESIDUAL_ANALYSIS, LINEAR_UNCERTAINTY, and MODEL_LINEARITY.

Introduction to UCODE_2005 Input and Output Files

The most commonly used UCODE_2005 input files are the main input file and the template, instruction, and derivatives-interface input files.

- The UCODE_2005 main input file is composed of data input blocks. For convenience, the input blocks may read data from other files. Each data block serves a specific purpose; for example, the Options and UCODE_Control_Data input blocks provide information about what is to be calculated by UCODE_2005.

- Template files are used to construct process-model input files using current parameter values. One or more template files are used for each UCODE_2005 run.

- Instruction files are used to extract, or read, values from process-model output files.

- The optional derivatives interface file enables UCODE_2005 to use sensitivities produced by other programs.

Keywords are used to identify most of the data in these files. This chapter refers to a few of the keywords to provide easy reference for users. All input blocks of the main input file, other input files, and keywords are described completely in Chapters 5 to 13 of this report.

The process model(s) executed by UCODE_2005 can include one process model or a sequence of models, and pre- and post-processors. These process models and processors need to be set up to run in batch mode.

The most commonly used UCODE_2005 output files are the main output file and data-exchange files. Here, the data-exchange files are described briefly.

Data-exchange files are computer data files produced by a computer program primarily for use by another program. The other program might be used by the modeler to generate graphs; for example, GW_CHART (Winston, 2000) or Microsoft's Excel. Alternatively, the program may use the data in calculations; for example, UCODE_2005 generates data-exchange files that are used by the other programs documented in this report.

Data-Exchange files contain data with little or no explanatory information. Most of the data-exchange files documented in this report contain one header line followed by columns of data. The header line contains labels for the columns of data in the file. Some of the columns of data contain identifying information such as an observation name or an integer value that can be used to control the symbol used in the graph. The data-exchange files are described in detail in Chapters 14 and 16 and listed in appendix B.

Data-exchange filenames begin with a prefix defined on the command lines, as discussed below. To make them distinctive, data-exchange filenames end with an extension that begins with an underscore. For example, ex1._ws is the name of a data-exchange file produced by an example distributed with UCODE_2005. The characters following the underscore reflect the file contents.

The data-exchange files are derived directly or in form from the JUPITER API (Appendix A).

Flowchart for UCODE_2005 Used to Estimate Parameters

A flowchart describing UCODE_2005 operation when it is used to estimate parameters is presented in figure 1.

Often parameters are estimated only after using starting parameter values to evaluate model fit and perform a sensitivity analysis to identify insensitive and correlated parameters. Execution of UCODE_2005 for these purposes proceeds through a subset of the steps used to estimate parameters.

Flowcharts for using UCODE_2005 with the other six codes documented in this report are presented in Chapters 15 and 17.

As shown in Figure 1, parameter-estimation begins by defining what is to be accomplished using data from the Options and UCODE_Control_Data input blocks. In the UCODE_Control_Data input block, keyword "Optimize" is used to indicate that parameters are to be estimated.

Next, process-model input files are created using the starting parameter values. This is accomplished by substituting the starting parameter values from the Parameter input blocks into the template files listed in the Model_Input_Files input block. UCODE_2005 then performs one execution of the process model(s) based on commands provided by the user in the Model_Command_Lines input block.

Next, for each observation, UCODE_2005 uses information from the Model_Output_Files input block and Instruction input files to read one or more values from the process-model output files. These values are used to calculate an equivalent simulated value to be compared to the observations defined in the Observation input blocks. Equivalent simulated values are called simulated values in the remainder of this discussion. Examples of calculating simulated values from values read from the output file(s) are described in Chapter 8. The simulated values calculated at this step of each parameter-estimation iteration are called unperturbed simulated values because they are calculated using the starting parameter values or, in the case of later iterations, the

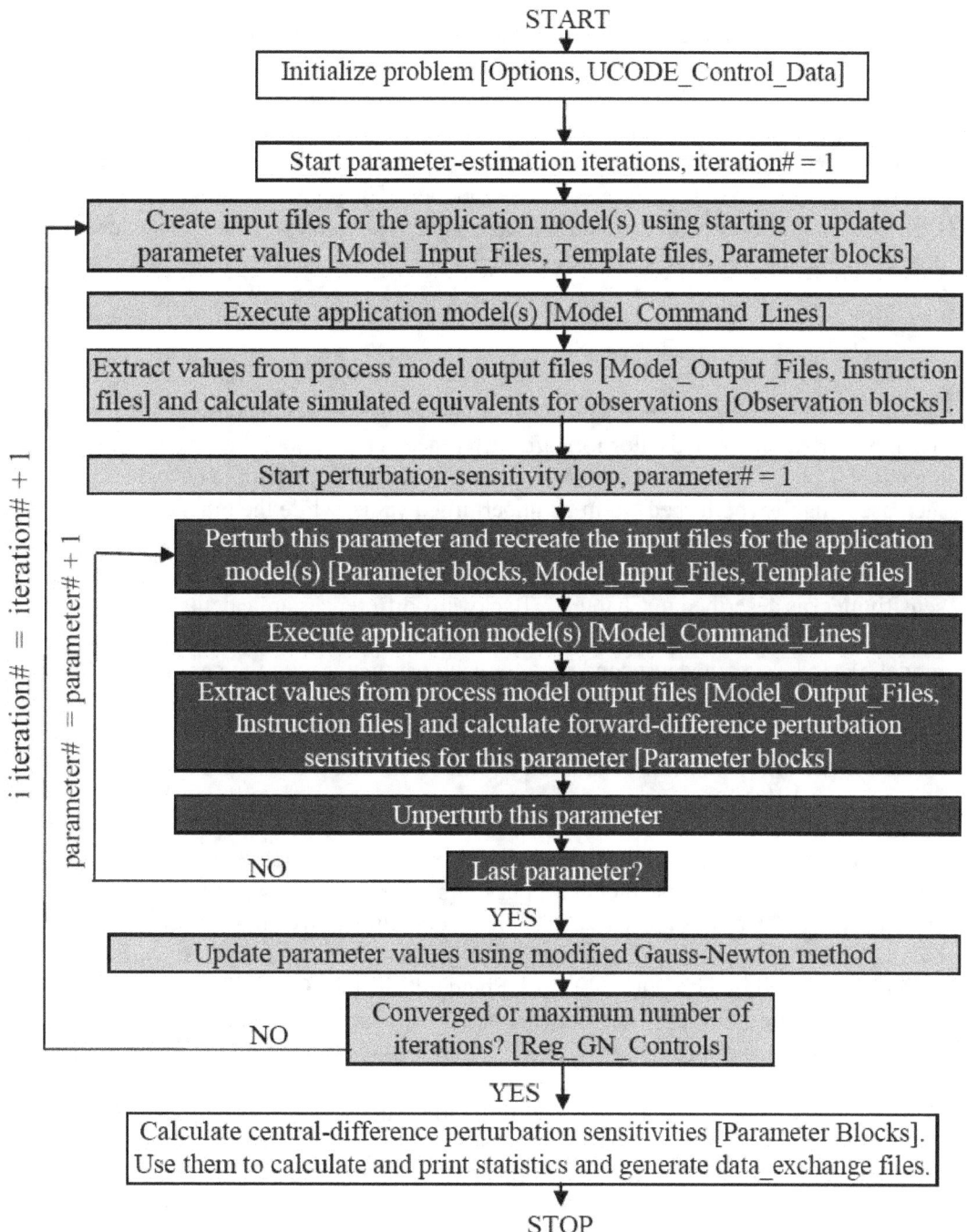

Figure 1: Flowchart showing major steps in the UCODE_2005 parameter-estimation mode using perturbation sensitivities. Selected input blocks of the UCODE_2005 main input file and other input files are listed in brackets; 'Parameter blocks' represents four input blocks used to define parameters, 'Observation blocks' represents three input blocks used to define observations. Iteration# is the parameter-estimation iteration number; parameter# is the parameter number. Gray shading is used to emphasize loops.

updated parameter estimates. Each observation minus the associated unperturbed simulated value is called a residual. The residuals are squared, weighted, and summed to produce the sum-of-squared-weighted residuals objective function, where the weighting is defined in the Observation input blocks. The objective function measures how well the model fits the observations. The goal of regression is to find parameter values that produce the smallest value of the objective function. The regression method in UCODE_2005 uses sensitivities to find those parameter values. (Hill and Tiedeman, in press, eq. 3-1 and 3-2; Hill, 1998, eq. 1 and 2).

Sensitivities are the derivatives of the simulated values with respect to the parameters, and can be obtained in two ways.

(1) As shown in Figure 1, observation sensitivities can be calculated by UCODE_2005 using perturbation methods. For forward-difference perturbation, the process model(s) are executed once for each parameter. For each execution, one parameter value is increased slightly (perturbed) from its unperturbed value, while the other parameter values are not perturbed. The differences between the resulting perturbed simulated values and the unperturbed simulated values are used to calculate forward-difference sensitivities, as described in Chapter 3. Backward-differences are calculated similarly except that the parameter values are decreased slightly. Alternatively, the process model(s) can be executed a second time for each parameter with the parameter values perturbed in the opposite direction and sensitivities can be calculated using more accurate central differences. The authors' experience indicates that this added accuracy is rarely needed to perform parameter-estimation iterations, though it is useful for calculating final statistics (see below).

 (2) Observation sensitivities can be calculated by process model(s) and read by UCODE_2005 from process-model output file(s). For example, when the Observation and Sensitivity Processes are active in MODFLOW-2000, unscaled sensitivities needed by UCODE_2005 can be printed to the MODFLOW-2000 output file with suffix _su which can be read by UCODE_2005 as a Standard File (see Chapter 11). For process-model sensitivities, the sensitivity loop Figure 1 is replaced by a single execution of the process model(s) and a Derivatives-Interface input file is needed.

A combination of process-model and perturbation sensitivities also can be used.

Once the residuals and the sensitivities are calculated, they are used to perform one parameter-estimation iteration. UCODE_2005 is distributed with the modified Gauss-Newton nonlinear regression parameter-estimation method described by Hill and Tiedeman (in press), which was modified from the method described by Hill (1998) and Cooley and Naff (1990). UCODE_2005 also provides the following additional ways to improve regression performance. First, dynamic omission of insensitive parameters can be used so that insensitive parameters do not disrupt regression performance. Second, unique criteria for each parameter can be specified that govern the maximum fractional amount that the parameter value can change in one parameter-estimation iteration. Smaller values may be useful for insensitive parameters. Third, a quasi-Newton or double-dogleg modification of the Gauss-Newton method can be used to reduce the number of parameter-estimation iterations needed and, in some cases, achieve successful regressions (Cooley and Hill, 1992; Dennis and Schnabel, 1996; Mehl and Hill, 2003).

The last step of each parameter-estimation iteration involves comparing two types of quantities against convergence criteria: (1) changes in the parameter values, where a unique criterion can be specified for each parameter, and (2) the change in the sum-of-squared-weighted residuals. If the changes are too large and the maximum number of parameter-estimation iterations has not been reached, the next parameter-estimation iteration is executed. If the changes are small enough, parameter estimation converges. If convergence is achieved because the changes in the parameter values are small enough (see 1 above), the parameter values are more likely to be the optimal parameter values – that is, the values that produce the best possible match between the simulated and observed values, as measured using the weighted least-squares objective function. If convergence is achieved because the changes in the objective function are small, it is less likely that the estimated parameters are optimal and, generally, further analysis is needed.

If parameter estimation does not converge and the maximum number of iterations has not been reached, the updated parameter values are substituted into the template files, and the next parameter-estimation iteration is performed.

When parameter estimation converges or the maximum number of iterations has been reached, sensitivities are calculated using the more accurate central-difference method. The additional accuracy is needed to achieve a sufficiently accurate parameter variance-covariance matrix (Hill and Tiedeman, in press, eq. 7-1; Hill, 1998, eq. 26), from which a number of useful statistics are calculated. If parameter-estimation converged, the final parameter values are considered to be optimized.

Brief Description of the Six Other Codes

The six other codes described in this documentation are
RESIDUAL_ANALYSIS,
RESIDUAL_ANALYSIS_ADV,
LINEAR_UNCERTAINTY,
MODEL_LINEARITY,
MODEL_LINEARITY_ADV, and
CORFAC_PLUS.

It is always useful to execute RESIDUAL_ANALYSIS when executing UCODE_2005. The analysis of model fit and leverage statistics provided are important to evaluating model fit to the observations. Often UCODE_2005 and RESIDUAL_ANALYSIS are both executed from one batch file.

The additional statistical and graphical analyses provided by RESIDUAL_ANALYSIS_ADV also is always useful, however, it can only be included in the batch file with the UCODE_2005 run under selected circumstances.

11

Predictions are often the ultimate focus of a modeling study. LINEAR_UNCERTAINTY can be used to calculate linear confidence and prediction intervals that approximate the uncertainty in predictions simulated using the process models and optimized parameter values. The likelihood that the intervals are affected by model nonlinearity can be evaluated using the post-processing program MODEL_LINEARITY and MODEL_LINEARITY_ADV.

It can be useful to calculate the predictions and their scaled sensitivities and linear confidence intervals throughout the calibration project to gain insight about how different conceptual models affect the predictions, the parameters important to the predictions, and the uncertainty of the predictions. In many circumstances the extra computational burden is minimal.

UCODE_2005 can also be used to calculate nonlinear confidence and predictions intervals, which quantify the uncertainty of model simulated values when the model is not linear. CORFAC_PLUS can be used to produce factors that allow intervals to account for model intrinsic nonlinearity and small-scale variations in system characteristics that are not explicitly accounted for in the model or the observation weighting.

Parallel-Processing Capabilities

UCODE_2005 is distributed with a parallel-processing capability that can substantially reduce execution times when calculating sensitivities or performing parameter estimation. In the flowchart shown in figure 1, the parallelization involves the sensitivity loop when sensitivities are calculated using perturbation methods; the process model runs are assigned to different processors for simultaneous execution. The parallel processing capability can take advantage of any number of processors linked using a local area network. The user needs to be able to access the computers and execute a program on them.

The parallel-processing capability is enabled in the executable file included in the UCODE_2005 distribution. To use this capability, two optional input blocks are needed in the main input file, as described Chapter 12 of this report.

Chapter 3: USER CONSIDERATIONS

Calibration of models of complex systems commonly is hampered by problems of parameter insensitivity and extreme correlation caused by data that are insufficient to estimate the parameters defined. Regression methods are imperfect tools that nevertheless can be very helpful in model calibration. To help modelers take advantage of these useful methods, this chapter provides a brief discussion of some key issues related to using sensitivity analysis and nonlinear regression methods to calibrate and analyze complex models. Chapter 16 of this report provides additional ideas in its discussion of using the output from UCODE_2005 and the three post-processors. More thorough discussions are provided by Hill and Tiedeman (in press) and Hill (1998).

The first section of this chapter lists a set of guidelines that can be thought of as organized common sense for model calibration with some new perspectives and statistics. The guidelines are discussed in detail in Hill and Tiedeman (in press); a previous version is in Hill (1998). The following sections discuss issues from the guidelines that are often of concern: parameterization, starting parameter values, perturbation sensitivity calculation and accuracy, weighting, sensitivity analysis, coping with a poorly posed model, alternative models, and residual analysis. A final section presents definitions of some terms related to confidence and prediction intervals.

Guidelines for Effective Model Calibration and Analysis using Nonlinear Regression

There are many opinions about how nonlinear regression can best be applied to the calibration of complex models, and there is not a single set of ideas that is applicable to all situations. It is useful, however, to consider one complete set of guidelines that incorporates many of the methods and statistics available in nonlinear regression, such as those suggested and explained by Hill and Tiedeman (in press) and listed in table 1. This approach has been used successfully even with exceptionally complex systems; for example, see D'Agnese and others (1997, 1999), Eberts and George (2000), and other reports listed in Chapter 15 of Hill and Tiedeman (in press). Table 1 is presented to introduce and remind the reader of the guidelines. Those who wish to use these guidelines are encouraged to read the complete discussion.

Parameterization

Parameterization is the process of identifying the aspects of the simulated system that are to be represented by estimated parameters. Most data sets only support the estimation of relatively few parameters. In most circumstances, it is useful to begin with simple models. Complexity can then gradually be incorporate as warranted by the complexity of the system, the inability of the model to match observed values, and the importance of the complexities to the predictions of interest (Guideline 1 of Table 1).

To obtain an accurate model and a tractable calibration problem, data not used directly as observations in the regression need to be incorporated into model construction (Guideline 2 of Table 1). For example, in ground-water systems, it is important to respect and use the known hydrogeology, and it is unacceptable to add features to the model to improve model fit if they contradict known hydrogeologic characteristics.

During calibration it may not be possible to estimate all parameters of interest using the available observations. In such circumstances, consider the suggestions of the section "Common Ways of Improving a Poor Model" in this report.

Table 1: Guidelines for effective model calibration (from Hill and Tiedeman, in press; modified from Hill, 1998).

Model Development
1. Apply the principle of parsimony (start simple; build complexity slowly)
2. Use a broad range of information to constrain the problem
3. Maintain a well-posed, comprehensive regression problem
4. Include many types of observations in the regression
5. Use prior information carefully
6. Assign weights that reflect errors
7. Encourage convergence by improving the model and evaluating the observations
8. Consider alternative models
Test the Model
9. Evaluate model fit
10. Evaluate optimized parameters
Potential New Data
11. Identify new data to improve model parameter estimates and distribution
12. Identify new data to improve predictions
Prediction Accuracy and Uncertainty
13. Evaluate prediction uncertainty and accuracy using deterministic methods
14. Quantify prediction uncertainty using statistical methods

Starting Parameter Values

Nonlinear regression begins with starting parameter values. There are three aspects of these starting values that are important.

1. In UCODE_2005, depending on the option chosen, the starting parameter values are used to calculate residuals, scaled and composite scaled sensitivities, and parameter correlation coefficients. These statistics are important for diagnosing potential problems with the model and the regression and for determining ways of addressing these problems. In most circumstances, it is useful to evaluate these statistics regularly as the model changes during the calibration process. The statistics printed by UCODE_2005 are discussed in Chapter 16 of this report, in Hill and Tiedeman (in press), and in part, in Hill (1998). Hill and Tiedeman (in press) include the equations

for the statistics used in this documentation and a discussion of how model nonlinearity affects the analysis.

2. It is sometimes advantageous to change the starting parameter values. As calibration proceeds, parameter values that produce a better model fit than the original starting parameter values are estimated by regression. Using the estimated parameter values to update the original starting parameter values used by UCODE_2005 in subsequent regression runs can reduce execution time because, commonly, fewer regression iterations are required when the starting parameter values produce a better model fit.

3. The starting parameter values can be used to test for the uniqueness of optimized parameter values; that is, the values at which the regression converges. This is accomplished by initiating the regression with different sets of starting values. If the sum of squared weighted residuals achieved is similar and resulting parameter estimates differ from each other by values that are small relative to their calculated standard deviations, the optimization is likely to be unique. If this is not the case, the optimal parameter values are not unique. Lack of uniqueness can be caused by a number of factors. If caused by local minima, it may be possible to examine the objective function value achieved by the different sets of parameter estimates and identify a global minimum as the set of estimated parameter values that is both reasonable and produces the smallest objective-function value. If non-uniqueness is caused by extreme parameter correlation, the objective-function value for each optimized set of parameters is likely to be similar and at least one pair of parameters will have a correlation coefficient very close to 1.0 or -1.0. This is demonstrated clearly by the simple test case presented by Poeter and Hill (1997). Difficulties with using correlation coefficients are discussed by Hill and Østerby (2003).

Perturbation Sensitivities

UCODE_2005 uses either sensitivities calculated by the process model or sensitivities calculated by UCODE_2005 using perturbation methods. This section describes how UCODE_2005 calculates perturbation sensitivities, how their accuracy can be improved, and how to address the common problem of calculated sensitivities being equal to zero when they should not be zero.

Calculation

A sensitivity equals the derivative of one simulated value with respect to one parameter, b. If the simulated value is the simulated equivalent to an observation, y', the sensitivity can be expressed as, $\partial y'/\partial b$. If the simulated value is a prediction, the sensitivity is generally expressed as $\partial z/\partial b$. For simplicity, sensitivities for simulated equivalents to observations are discussed in the rest of this section. However, the comments apply to sensitivities for predictions as well.

Sensitivities serve two functions in inverse modeling. First, they are useful indicators of both the importance of the observation to the estimation of the different parameters and the importance of each parameter to the simulated values. Second, they are needed by the modified Gauss-Newton method to determine parameter values that produce the best fit, as measured by a weighted least-squares objective function. For additional discussion about sensitivities and their utility, see Hill and Tiedeman (in press, Chapter 4 and Guideline 3) or Hill (1998, eq. 8-13 and Guideline 3).

In UCODE_2005, sensitivities can be calculated approximately using either a forward-, backward-, or central-difference approximation. For forward differences, each sensitivity (one for each observation with respect to each parameter) is calculated as:

$$\frac{\Delta y'}{\Delta b} = \frac{y'(\mathbf{b} + \Delta \mathbf{b}) - y'(\mathbf{b})}{(\mathbf{b} + \Delta \mathbf{b}) - (\mathbf{b})} \qquad (1)$$

where:

$\mathbf{b}$ a vector (can be thought of as a list) of the estimated parameter values;

$y'(\mathbf{b})$ the value of the simulated value, y', calculated using the parameter values in $\mathbf{b}$;

$\Delta \mathbf{b}$ a vector in which all values are zero except for the one that corresponds to the parameter for which sensitivities are being calculated;

$y'(\mathbf{b} + \Delta \mathbf{b})$ the value of y' calculated using the parameter values in $(\mathbf{b} + \Delta \mathbf{b})$;

$\Delta y'$ the change in the simulated value caused by the parameter value change, Δb;

Δb the nonzero value in $\Delta \mathbf{b}$, which is called the perturbation for this parameter;

The derivative is said to be "evaluated for the parameter values in $\mathbf{b}$". For nonlinear problems, this is important because the sensitivities can be different for different values in $\mathbf{b}$.

The size of the perturbation, Δb, is calculated as a user-specified factor (PerturbAmt) times the unperturbed parameter value. To calculate backward instead of forward differences, specify a negative factor. If the unperturbed-parameter value equals zero during the regression, the perturbation is calculated using the starting parameter value. If the starting parameter value equals zero, a value of 1.0 is used to calculate Δb, so that Δb = PerturbAmt in that circumstance. Calculating the sensitivities for each parameter using either forward or backward differences requires that the process model be run once for the unperturbed parameters and an additional time for each parameter being adjusted. The flowchart of figure 1 includes the steps by which forward-difference sensitivities are calculated.

Central-difference sensitivities are more accurate than forward-difference sensitivities, but require two runs of the process model(s): in one the perturbed parameter is increased, in one it is decreased. Execution time is thus increased by about a factor of two. The central-difference sensitivities are calculated as:

$$\frac{\Delta_2 y'}{\Delta_2 b} = \frac{y'(\mathbf{b} + \Delta \mathbf{b}) - y'(\mathbf{b} - \Delta \mathbf{b})}{(\mathbf{b} + \Delta \mathbf{b}) - (\mathbf{b} - \Delta \mathbf{b})} \qquad (2)$$

where Δ_2 is used to denote the central-difference. Again, the derivative is said to be "evaluated for the parameter values in **b**". The added accuracy of the central-difference approximation is needed when the variance-covariance matrix is calculated. It rarely is needed for the regression if the suggestions described in the following section are followed, but UCODE_2005 allows use of central-difference sensitivities in the regression if selected by the user.

Accuracy

The accuracy of the perturbation sensitivities calculated by UCODE_2005 depends on the precision of the extracted simulated values, the magnitude of the simulated values, and the number of digits used for the substituted parameter values.

For example, consider the problem in Appendix C, which uses the process model MODFLOW-2000 (Harbaugh and others, 2000) with the preconditioned conjugate-gradient (PCG2) solver (Hill, 1990) with HCLOSE=$1x10^{-5}$ and RCLOSE=$1x10^{-5}$. For this problem, with hydraulic heads typically in the 100's, heads from this partially double-precision model are expected to be accurate to about seven significant digits (four to the right of the decimal point). The _os file produced by MODFLOW2000 provides seven significant digits for all simulated equivalents to observations, which reflects the likely accuracy of the simulated heads. It is probably more digits than are warranted for the simulated drawdowns and streamflow gains, which are calculated by subtracting heads and flows. However, the additional digits do not tend to deteriorate the accuracy of the perturbation sensitivities.

The parameter values substituted into the template files have fifteen or 16 significant digits, depending on the value. It is important to have sufficient digits, and this certainly is enough. Sensitivities calculated by UCODE_2005 using forward differences with PerturbAmt = 0.01, or one percent of the parameter value, and central differences with PerturbAmt=0.01 and 0.001 were compared with the more accurate sensitivity-equation sensitivities calculated by the Sensitivity Process of MODFLOW-2000 (Hill and others, 2000) using the PCG2 solver with HCLOSE = RCLOSE = $1x10^{-5}$. For this situation, the forward sensitivities were accurate to about three digits to the right of the decimal point. The central difference sensitivities with PerturbAmt=0.01 were a bit better, and those with PerturbAmt=0.001 were worse. Such differences often do not affect estimated parameter values. This is demonstrated by several of the problems considered by Yager (2004). Problems for which there was a difference tended to be very nonlinear and in such cases perturbation sensitivities sometimes produced better results. Inaccurate sensitivities can produce enough error in calculated parameter correlation coefficients (Hill and Østerby, 2003) to make these statistics unreliable indicators of extreme parameter correlation.

The accuracy of the perturbation sensitivities sometimes can be improved by careful consideration of the number of significant figures used for both the simulated values and the parameter substitution. The simulated values and parameter values of equations 1 and 2 need sufficient precision to maintain a reasonable number of significant digits after the subtraction. Use of more digits than are precise given the process model, however, may not improve accuracy. Given the use of single and double precision variables in the process model and the accuracy of the computer, the additional figures may be meaningless, in which case accuracy of the perturbation sensitivities would not be improved.

Opportunities sometimes exist to improve the accuracy of the simulation. In ground-water flow problems, for example, one such situation occurs when the simulated hydraulic heads are in the 100's or 1000's. Often a simple change in datum results in simulated values being consistently in the 10's, which allows more of the available significant figures to be used to improve the accuracy of the solution. For example, the accuracy of the problem in Appendix C could be improved by raising the datum by 100m. To accomplish this, 100 m would need to be subtracted from the hydraulic-head observations, the river elevation, and the elevation of the layers. That example also might benefit from using more significant digits for the substituted parameter values. In transport simulations, similar improvements may be attainable by scaling, log-transforming, or using special weighting schemes for the concentrations (For example, see Barth and Hill, 2005, in press). Often, precision is improved by log-transforming parameters used to calculate hydraulic-conductivity and storage properties.

For nonlinear parameters, the accuracy of the sensitivities also depends on the size of the parameter perturbations. Determining the appropriate size can be problematic. Theoretically, the perturbation sensitivities approach the exact sensitivities as the perturbation size decreases. However, perturbations that are too small can result in negligible differences in the simulated values or differences that are obscured by round-off error. A perturbation that is too large, however, can yield inaccurate sensitivities for nonlinear parameters because it does not capture the slope of the function at a specific set of parameter values. The user needs to be aware of the potential difficulties and may need to experiment with different perturbation sizes.

Even in nonlinear problems, the sensitivities for some parameters can be linear. That is, the same sensitivity is calculated for all perturbation amounts large enough to produce sufficient significant digits in the simulated values, given that the values of the other parameters do not change. For example, in ground-water problems, recharge parameters have linear sensitivities if the forward problem is linear (the system is confined and all boundary conditions are linear). For such parameters, perturbations generally can be large, and inaccurate sensitivities result only when the perturbation is so small that the numerators of equations 1 and 2 are dominated by round-off error.

Weighting Observations and Prior Information

Observations and prior information need to be weighted so that (1) the weighted residuals are all in the same units so that they can be squared and summed in the least-squares objective function and (2) to reflect the relative accuracy of the measurements (Hill and Tiedeman, in press, Chapter 3; Hill, 1998, p. 4, 13-14, 45). For a valid regression, weighting needs to be proportional to the inverse of the variance-covariance matrix of the errors in observations and prior information [Draper and Smith, 1998, p. 222]. Hill and

Tiedeman (in press) and Hill (1998) suggest that the user attempt to define weighting that **equals** the inverse of the error variance-covariance matrix. The following discussion presumes that the weighting follows this suggestion.

When the observation or prior information errors are independent of one another, the weight matrix is diagonal. Each non-zero element of the diagonal weight matrix equals one over the variance of the error. Variances are rarely easy for users to understand, so UCODE_2005 allows users to specify one of the following: variance, standard deviation, or coefficient of variation. The weight or square-root of the weight also can be specified. This allows the statistic that makes most sense in a given situation to be used in the input file. For example, streamflow observation error may be most readily understood based on a percent of the observed value, which can be most easily expressed as a coefficient of variation of, for example, 0.20, or 20 percent. Hydraulic-head observation error is more often understood as some number of feet, meters, or centimeters, and is most easily expressed as a standard deviation of, for example, 0.3 meters. More detailed information about determining values for weights is provided in Hill and Tiedeman (in press, Guideline 6) or Hill (1998, p. 46-49).

When a coefficient of variation is specified, the weight ω_i ideally would be calculated using the true value of the observed quantity or prior information, so that,

$$\omega_{i_i} = \frac{1}{\left(cv_i \tilde{y}_i + \eta\right)^2} \qquad (3)$$

where i identifies the observation or prior information, cv_i is the coefficient of variation, $\tilde{y}_i$ is the true value, and η is a constant that can be added to control how much small values of $\tilde{y}_i$ influence the results. The value of $\tilde{y}_i$ is, in general, unknown and is approximated using either the observed value [for example, Keidser and Rosbjerg, 1991] or the simulated value [for example, Wagner and Gorelick, 1986; Barlebo and others, 1998]. Anderman and Hill (1999) compare the two approaches and show that using observed values tends to produce biased parameter estimates. In UCODE_2005 an observation for which the coefficient of variation is used to define the weighting can either (1) use observed values in equation 3 and η=0.0 or (2) use observed values at first and use simulated values as regression approaches a solution, with η specified by the user.

One approach is to divide the objective function into terms that include subsets of the observations and prior information. For each subset, the weighting can be conceptualized as a multiplicative factor times a matrix (Gailey and others, 1991; Barlebo and others, 1998). UCODE_2005 provides the ability to define a weight multiplication factor for any user-defined group of observations or prior information.

In UCODE_2005, special functionality is provided if the objective function is divided into two terms: a term that includes observations and prior information with weights that are never calculated with simulated values and a term that includes observations with weights that can be calculated with simulated values. In ground-water models, the first term often includes head observations and the second term often includes concentration and flow observations. This produces equations 4 to 6.

The variance-covariance matrices for the errors in observations and prior information for the two terms of the objective function become

$$V_1 = \sigma_1^2 U_1$$
$$V_2 = \sigma_2^2 U_2 \tag{4}$$

The subscripts 1 and 2 indicate the two terms of the objective function. V_1 and V_2 are the variance-covariance matrices of the errors associated with the two sets of data, σ_1^2 and σ_2^2 are the multiplicative factors, and U_1 and U_2 are matrices. If U_1 and U_2 are defined to approximate V_1 and V_2, the expected value of σ_1^2 and σ_2^2 is 1.0. The objective function, $S(\underline{b})$, can be expressed as

$$S(\underline{b}) = (1/\sigma_1^2) S_1(b) + (1/\sigma_2^2) S_2(b) \tag{5}$$

Typically, this equation is rearranged to

$$S'(b) = S_1(b) + (\sigma_1^2/\sigma_2^2) S_2(b) = S_1(b) + \lambda_r S_2(b)$$

where λ_r is called the scaling factor. To adjust the weighting to ensure the desirable result that the different subsets of data have, on average, equal variance, the scaling factor can be calculated as:

$$\lambda_r = (\sigma_1^2/\sigma_2^2) = \frac{\dfrac{S_1(\underline{b}_r)}{n_1}}{\dfrac{S_2(\underline{b}_r)}{n_2}} \tag{6}$$

The subscript r identifies the parameter-estimation iteration. Variables with this subscript change each parameter-estimation iteration. n_1 is the number of items in the first term of the objective function and n_2 is the number in the second term. Care needs to be taken that the resulting weighting is reasonable (Guideline 6; Hill and Tiedeman, in press; Hill, 1998). This is aided by defining U_1 and U_2 to approximate V_1 and V_2. In this case, λ_r values that differ too much from 1.0 indicate that the final weighting may be unrealistic. If any weights are calculated using simulated values, the scaling factor is initially set to 1.0 and is subsequently calculated with equation 6 at each parameter-estimation iteration.

The weight matrix is full when there are non-zero covariances, which are the off-diagonals of the error variance-covariance matrix (Hill and Tiedeman, in press, eq. 3-2; Hill, 1998, p.7, eq. 2). A diagonal weight matrix is strictly valid only if the measurement errors are independent. UCODE_2005 can accommodate a diagonal or full weight matrix for any set of observations or prior information, except for groups for which the weighting is defined using simulated values. Thus, in equation 4, U_1 can be a full matrix;

U2 is always a diagonal matrix. There are two issues to consider related to full weight matrices. The first is that many common types of correlated errors can be addressed by differencing of the data. The second is whether a full weight matrix is important in practice.

In some circumstances differencing methods can be used to address commonly encountered error correlation so that a diagonal weight matrix applies. For example, hydraulic heads measured over time at a single well that has a poorly determined elevation can be represented as changes from the first measured head. A second example is for heads measured at wells that have been surveyed relative to one another. One well can be defined as the reference and that value subtracted from the others. In both situations, the error in elevation is eliminated using differencing. For more discussion of differencing, see Hill and Tiedeman (in press, guideline 6) and Hill (1998, guideline 6).

The importance of using a full weight matrix even in the presence of correlated measurement errors is questionable. A published study by Christensen and others (1998) and unpublished numerical investigations by Mary C. Hill (U.S. Geological Survey, written communication, 1996) indicate that typical error correlations have little effect on nonlinear regression, residual analysis, or uncertainty analysis. This, however, is a preliminary conclusion drawn from partial, limited investigations. Further work remains to determine the importance of using full weight matrices in problems typical of ground-water investigations. It is hoped that the ability of UCODE_2005 to represent full weight matrices will enhance understanding of this issue.

Sensitivity Analysis

Sensitivity analysis is used to explore the relations between observations, parameters, and predictions produced by the constructed model. As discussed by Saltelli and others (2000) and Hill and Tiedeman (in press), sensitivity analyses can be local, using sensitivities calculated for a given set of parameter values. Sensitivity analysis can also be global, using values simulated using many sets of parameter values. UCODE_2005 and the three postprocessors include a number of capabilities for local sensitivity analysis and one method for global sensitivity analysis, as shown in Table 2.

Table 2. Statistics for sensitivity analysis provided in UCODE_2005 and the other six
programs documented in this report.

Relation explored	Statistics
Local sensitivity analysis	
Observations-Parameters	Dimensionless scaled sensitivities (dss)
	Composite scaled sensitivities (css)
	Parameter correlation coefficients (pcc)
	Leverage
	Parameter confidence intervals
	Cook's D
	DFBetas
Parameters-Predictions	Prediction scaled sensitivities (pss)
Observations-Parameters-Predictions	Prediction confidence intervals
Global sensitivity analysis	
Observations-Parameters	Results of investigate-objective-function mode

Local sensitivity analysis has the advantage of requiring much less execution time than
global sensitivity analysis. Local sensitivity analysis is often conducted routinely during
model calibration to identify insensitive and extremely correlated parameters. For
example, during calibration, parameter correlation coefficients can be used to indicate
whether the observations provide enough information to estimate each of the defined
parameters. This can then be used to decide which parameters to estimate, as discussed
by Poeter and Hill (1997) and Hill and Østerby (2003). When considering an optimal set
of parameter values, parameter correlation coefficients can be used to determine if
optimized parameter values are likely to be unique.

Scaled sensitivities are useful measures of the information provided by observations for
many types of parameters, but not for parameters such as the hydraulic head at constant-
head boundaries that would change with a datum change (Hill and Tiedeman, in press,
Chapter 4.4.1).

UCODE_2005 provides optional dynamic omission of insensitive parameters to improve
regression performance in the presence of insensitive parameters. The method is
described in the input instructions.

Common Ways of Improving a Poor Model

Problems such as insensitivity and extreme correlation of parameters and poor model fit
are common in model calibration. Possible ways of addressing these problems are as
follows, listed in order of how often the suggestion is most appropriate in practice.

1. Reconsider the model construction, including geometry of internal and external
   boundaries, discretization, and so on. For example, in ground-water models, consider
   the internal definition of hydrogeologic units. Regression difficulties and poor model
   fit can help reveal misconceptions used to construct the model and mistakes in input
   data.

2. Modify the defined parameters by adding, omitting, and (or) linking parameters to be estimated. See section "Parameterization" above. The procedure for linking parameters is explained in the input instructions presented in this report.
3. Carefully eliminate observations or prior information if available evidence indicates that they are likely to be biased. Do not omit observations just because the model does not fit them well.
4. Adjust weights either for groups of observations and prior information, or perhaps individually. Small changes in the weighting rarely affect regression results, so, in most circumstances, avoid time-consuming repeated runs using slightly different weights.

A useful approach is to continually strive to identify and correct inaccuracies in the model construction or the use of observations (this is guideline 7 of table 3). Use the model fit and calculated parameter sensitivities and correlation coefficients to facilitate this process. Nearly always, nonlinear regression will converge as the problems are resolved. Additional potential difficulties and their resolutions also are discussed in Hill and Tiedeman (in press).

Alternative Models

The sparse data sets available for the development of most environmental models often support feasible alternative conceptual models. Feasible conceptual models are those that reasonably represent known conditions and yield an acceptable fit to the data with reasonable parameter values. It is important to evaluate models that represent as many potentially important undefined aspects of the system as possible.

Feasible alternative models need to be used to make predictions and to determine the associated confidence in those predictions. Multi-model averaging facilitates evaluation of predictions and their associated uncertainty (Poeter and Anderson, 2005; Poeter and others, written commun., 2005). If the various models produce confidence intervals on predictions that are so large that the appropriate scientific conclusion or management decision is unclear and additional data collection is warranted, statistics of the regression can be used to help identify new data that are most likely to reduce the uncertainty and differentiate the models. This can facilitate development of models that are more representative of the system (Hill and others, 2001; Tiedeman and others, 2003, 2004; Matthew Tonkin, C.R. Tiedeman, M.C. Hill, and D.M. Ely, written commun., 2005).

Residual Analysis

To judge whether a model represents a system accurately, it is crucial to analyze the residuals (observed minus simulated values). A complete analysis of residuals includes consideration of summary statistics and consideration of graphs and maps of weighted and unweighted residuals (see of Hill and Tiedeman, in press, Chapter 6 and Guideline 9; and Hill, 1998, guideline 7). In the graphical analyses, some departure from ideal patterns may be attributed to the limited number of data and the fitting of the regression. The effect of these contributions can be evaluated by generating random data sets that have the same number of data and characteristics consistent with the fitting of the regression (Cooley and Naff, 1990; Christensen and Cooley, 2005). Such random data sets can be generated with UCODE_2005 output files, an optional user-generated input file, and the computer programs RESIDUAL_ANALYSIS and RESIDUAL_ANALYSIS_ADV, which are documented in Chapters 14, 15, and 17 of this report.

Predictions and Their Linear Confidence and Prediction Intervals

Often models are constructed to make predictions of what is likely to occur under different circumstances. Predictions for these conditions can be simulated using a calibrated model. Generally the uncertainty of the predictions also is of interest. This section addresses two issues related to predictions and prediction uncertainty: (1) The use of differences between predictions and (2) the different types of intervals that can be used to represent prediction uncertainty.

Differences between predictions often are of interest. For example, two remediation scenarios may be evaluated based on the difference in concentration simulated at a supply well. Of concern is whether one is really better than the other, which can be evaluated be determining whether the difference between predicted concentrations is significantly different than 0.0. With UCODE_2005, the derived prediction capability and the ability to run multiple process models using batch files make it easy to consider predicted differences that are calculated by subtracting values produced by a base simulation from values produced by a predictive simulation. That is:

(value from predictive simulation) - (value from base simulation) = difference.   (7)

The base simulation may represent conditions related to the calibration or to one of several future scenarios being considered. In a ground-water example, values of interest might be hydraulic heads at the same location before and after additional pumpage is imposed on the system. In this circumstance, the predictive simulation includes the additional pumpage; the base simulation does not. The difference would be the drawdown resulting from the pumpage. The use of differences is discussed further by Hill and Tiedeman (in press, Section 8.4.5) and Hill (1994).

UCODE_2005, used in conjunction with post-processor LINEAR_UNCERTAINTY, includes linear methods of calculating and evaluating predictions. This section introduces those methods. Detailed information about LINEAR_UNCERTAINTY and how to use UCODE_2005 and LINEAR_UNCERTAINTY is provided by Hill and Tiedeman (in press), Hill (1994), and Chapter 15 of this report.

The program LINEAR_UNCERTAINTY calculates 95-percent linear confidence and prediction intervals on predictions using equations shown in Hill and Tiedeman (in press, Chapter 8.3) and Hill (1998, eq. 28). UCODE_2005's Nonlinear-Uncertainty mode can be used to calculate nonlinear confidence and prediction intervals. Confidence and prediction intervals can be defined as follows:

Confidence intervals represent the uncertainty in the simulated values that results from the uncertainty in the parameter values. For linear confidence intervals, the uncertainty in the parameter values is expressed by the parameter variance-covariance matrix (Hill and Tiedeman, in press, eq. 7-1; Hill, 1998, eq. 28). The validity of linear confidence intervals depends on the calibrated model accurately representing important aspects of the true system, the model being linear, and the weighted residuals being normally distributed. For nonlinear intervals, the uncertainty in the parameter values is expressed by the parameter confidence region. The validity of nonlinear intervals depends on the calibrated model accurately representing important aspects of the true system

Prediction intervals include the uncertainty in the parameter values as described for confidence intervals, and also include the effects of the measurement error that is likely to be incurred if the predicted quantity is to be measured. Prediction intervals are larger than confidence intervals and need to be used when a measured value is to be compared to a calculated interval.

Whether confidence or prediction intervals are used depends on whether or not the effects of measurement error are to be included. The idea of a prediction interval is distinct from the predictions, but the identical terminology can cause confusion. This terminology is firmly entrenched in statistics; here we suggest careful use of the terms prediction and prediction interval.

There are several ways to calculate confidence and prediction intervals, depending on how many predictions are to be considered. The methods fall into two categories: individual and simultaneous intervals. These are defined as follows.

Individual intervals apply when only one prediction is of concern. There is only one method of calculating individual linear confidence and prediction intervals (Hill and Tiedeman, in press, eq. 8-12 and 8-13; Hill, 1994, eq. 11 and 15), and it is exact if the model is linear and accurate, and the residuals are normally distributed.

Simultaneous intervals apply when the number of predictions of concern exceeds one, or when the interval is calculated on a quantity that is not precisely defined, such as the largest value wherever it occurs within the model.

Different types of simultaneous intervals are appropriate for different circumstances. The names of the possible intervals are "Bonferroni", "Scheffé d=k", and "Scheffé d=np", and all are approximate. If the number of predictions (represented by k) exceeds one and is less than the number of parameters, np, either approximate Bonferroni or Scheffé d=k simultaneous intervals apply. If k is greater than np, Scheffé d=np simultaneous intervals apply. Both the Bonferroni and Scheffé d=k methods tend to produce intervals that are larger than exact intervals would be for a linear, accurate model with normally distributed residuals. Therefore, the smaller of the two intervals needs to be used. LINEAR_UNCERTAINTY only prints the smaller of the two intervals.

If the number of predictions of concern cannot be exactly defined, simultaneous linear confidence and prediction intervals using the approximate Scheffé d=np method apply.

Scheffé d=np intervals tend to be larger than exact linear intervals would be for a linear, accurate model calculated for the same circumstances.

Calculating the different types of intervals depends on whether linear or nonlinear intervals are calculated. Within each of those two categories, the methods differ only in the critical values used (Hill and Tiedeman, in press, Chapter 8.4; Hill, 1994, eq. 11-17). The critical values are statistics from standard probability distributions. The probability distributions of concern are the Student-t, Bonferroni-t, and F-distributions. Tables of the statistics from these distributions were programmed into LINEAR_UNCERTAINTY and the nonlinear-uncertainty mode of UCODE_2005. The appropriate critical value is determined by the program based on information provided by the user. Two types of intervals are considered -- individual and simultaneous. There are three ways of calculating linear simultaneous intervals. Because the calculations are quick, LINEAR_UNCERTAINTY calculates all of the intervals and prints three of them after eliminating one of the simultaneous intervals because it is less accurate than its alternative, as discussed in Chapter 15. Of the three intervals printed, the user needs to choose the appropriate interval for a given application. The nonlinear-uncertainty mode calculates and prints a single type of interval, as indicated by the user.

The uncertainty analysis can include only the estimated parameters, or can also include parameters that were not estimated because of insensitivity, parameter correlation, or both. Including unestimated parameters can be important if the unestimated parameter values are important to the predictions. Including unestimated parameters is discussed further in Chapter 15 and in Hill and Tiedeman (in press, Sections 7.2.5 and 8.1). The example presented in Appendix C includes the effects of an unestimated parameter on measures of prediction uncertainty.

Linear confidence and prediction intervals can be useful indicators of prediction uncertainty (Christensen and Cooley, 1999; Hill and Tiedeman, in press, Chapter 8, exercise 14). However, as mentioned above, their utility depends on model linearity and the model adequately representing important aspects of the system. As these criteria are violated, the stated significance level of the intervals becomes increasingly questionable. Instead of a 95-percent interval, for example, the interval may in reality reflect a 99- or 50-percent significance level. These differences can have considerable consequences. For example, if a remediation effort is to be designed to accommodate a 95-percent uncertainty level, designing it instead to a 99-percent uncertainty level requires considerable more expense. Designing it to a 50-percent uncertainty level would produce an unacceptable level of risk. Model linearity can be tested with the UCODE_2005 post processor MODEL_LINEARITY, while model accuracy can be evaluated by analyzing model fit as mentioned in the earlier section "Residual Analysis." The proper use (and potential inaccuracies) of using linear confidence and prediction intervals, for nonlinear problems, are discussed by Hill and Tiedeman (in press), Christensen and Cooley (1999) and Hill (1994, 1998). Using nonlinear intervals removes the requirement of the model being linear.

Other common problems occur when the predictions of interest include types of quantities not included in the observations used to calibrate the model, the prediction conditions differ dramatically from the calibration conditions, or aspects of the system not represented by defined parameters contribute significant uncertainty. In such a circumstance, confidence and prediction intervals may not accurately indicate prediction uncertainty and need to be used with caution. In some cases including unestimated parameter values in the analysis may be useful, but this option has not been researched.

When a number of alternative models are considered and calibrated, multi-model averaging facilitates evaluation of predictions and their associated uncertainty (Poeter and Anderson, 2005). The computer code MMRI (Multi-Model Ranking and Inference, E.P. Poeter, M.C. Hill, E.R. Banta, written commun., 2005) can be used to calculate model-averaged predictions and uncertainties using data-exchange files from UCODE_2005.

Chapter 4: RUNNING UCODE_2005, RESIDUAL_ANALYSIS, MODEL_LINEARITY, and LINEAR_UNCERTAINTY

Running UCODE_2005

The Run Command for UCODE_2005 needs to be executed from the directory containing the UCODE_2005 input files. The process model input, output, and batch files can be in one or more separate directories. The process model(s) need to execute completely from one batch file without human intervention. The batch file can in turn run other batch files, and in this way there can be multiple process models.

The UCODE_2005 run command is of the form:

path:\UCODE_2005 input-file fn

where:

path:\ = the relative or absolute path to the UCODE_2005.exe on your computer (alternatively you could specify this in your system path variable)

input-file = the name of the main UCODE_2005 input file (these files have extension 'in' in the examples distributed; see appendices C and D)

fn = filename prefix for UCODE_2005 output files (spaces are not allowed in fn, even on operating systems that allow spaces in filenames)

Controlling Execution and Output

Table 3 lists the modes in which UCODE_2005 can be executed, the input block keywords that control the modes, and the major output of interest to most users. A few comments about each mode are provided here; the rest of the document provides extensive information on data input, execution, and UCODE_2005 output for each mode.

The **forward mode** is generally run first to check for errors in model construction, path specification, and calculation of simulated values.

The **sensitivity-analysis mode** produces statistics calculated for the parameter values specified in the UCODE_2005 main input file. The statistics calculated include scaled sensitivities and, depending on the value of SenMethod, parameter correlation coefficients and parameter confidence intervals. These can be used for sensitivity analysis as described in Chapter 3 and references cited therein.

The **parameter-estimation mode** is used to calculate parameter values that provide a better fit to the objective function.

Table 3: Modes of UCODE_2005, the source of the parameter values, commonly used model output, and input block keywords that control the mode.
[fn, replaced by the prefix listed on the command line (Chapter 4); OPT, Options input block (Chapter 6); CON, UCODE_Control_Data input block (Chapter 6); MOD, Model_Command_Lines input block (Chapter 6); PAR, Parameter_Groups, Parameter_Data, Parameter_Values input blocks via template files (Chapters 7 and 11); * is replaced by 'conf' or 'pred', see Chapter 17.]

Mode[1]	Source of Parameter Values	Name of Main Output File and Commonly Used Model Output [2]	Input Block Keywords[3]
The first five modes are typically executed in order. The last two of these five need to follow a successful Parameter-Estimation mode run. See section later in Chapter 4.			
Forward	PAR	Main output file: fn.#uout • Fit of simulated equivalents to observations (_os, _r, _w, _ws)	OPT: Verbose=4 or 5 to check simulated equivalents MOD: Purpose=forward
Sensitivity-analysis	PAR	Main output file: fn.#uout • As for forward mode • Composite and dimensionless scaled sensitivities (_sc, _sd) • Parameter correlation coefficients and confidence intervals (_pcc, _pc)[5]	CON: Sensitivities=yes [4,5]PAR: SenMethod=-1,0,or 2 [4]MOD: Purpose=forward or forward&der
Parameter-estimation	PAR or values calculated by regression	Main output file: fn.#uout • Optimal parameter values or data from parameter-estimation iterations to diagnose problems. (_pa, _pc, _ss, _pe)	CON: Sensitivities=yes Optimize=yes [4]PAR: SenMethod=-1,0,1,or 2 [4]MOD: Purpose=forward or forward&der
Test-model-linearity	Data-exchange file fn._b1	Main output file: fn.#umodlin • Simulated values for Beale's measure. (_b2)	CON: Linearity=yes [4]MOD: Purpose=forward or forward&der
Prediction	Data-exchange file _paopt	Main output file: fn.#upred • Predictions and prediction scaled sensitivities (_p, _spsr, _spsp, _sppr, _sppp)	CON: Sensitivities=yes Prediction=yes [4]PAR: SenMethod=-1,0,1,or 2 [4]MOD: Purpose=forward or forward&der
The next two modes need to be coordinated with other runs. See Chapter 17.			
Advanced-test-model-linearity	fn._b1adv*	Main output file: fn.#umodlinadv_* • Simulated values linearity measures. (_b2adv*, _b4*)	CON: LinearityAdv=* [4]MOD: Purpose=forward or forward&der
Nonlinear-uncer-tainty	PAR or values calculated by regression	Main output file: fn.#unonlinint_* • (_int*, _int*par, _int*wr)	CON: NonLinearIntervals=yes MOD: Purpose=forward or forward&der
The last mode is independent of all other runs.			
Investigate-objective-function	PAR via template files	Main output file: fn.#usos • Parameter values and associated objective-function values (_sos)	CON: SOSSurface=yes MOD: Purpose=forward

[1] **Forward**: the process model is executed once. **Sensitivity-analysis**: for perturbation sensitivities, the process model is executed once or twice for each parameter (SenMethod=1 or 2); for process-model sensitivities, the process model is executed once (SenMethod=-1 or 0).

Parameter-estimation: process model is executed as for sensitivity-analysis for each parameter-estimation iteration plus one (the last is for final statistics). **Prediction**: as for sensitivity-analysis; use Sensitivities=no to check calculation of predictions. **Test-model-linearity** and **Advanced-test-model-linearity**: the process model is executed twice for each parameter. **Nonlinear-uncertainty**: Nonlinear regression is executed once for each limit of each interval. **Investigate-objective-function**: the process model is executed once for each set of parameter values.

[2] Selected data-exchange file extensions are listed in parentheses. See Chapter 14 and 17.

[3] Verbose, Sensitivities, Optimize, SenMethod, Purpose, Prediction, Linearity, and SOSSurface are keywords of the noted input block. See input block descriptions for more information. Set keywords Sensitivities, Optimize, Linearity, Prediction, LinearityAdv, NonlinearIntervals, and SOSSurface to 'no' by default or designation, except as noted.

[4] If SenMethod=-1 or 0, use Purpose=forward&der. Purpose=derivative exists but is rarely used.

[5] Parameter confidence intervals and parameter correlation coefficients (pcc) are not calculated if SenMethod=1, so that option is not listed.

The **test-model-linearity mode** uses parameter values generated in the parameter-estimation mode and stored in the data-exchange file with extension _b1. It produces a data-exchange file with extension _b2. The latter is needed by MODEL_LINEARITY to test the accuracy of linear confidence intervals on parameters and predictions.

The **prediction mode** is used to calculate predictions and associated sensitivities. Results are used by LINEAR_UNCERTAINTY. The conditions simulated by the process model may be different for the prediction mode than for previously defined models.

The **advanced-test-model-linearity mode** conducts additional tests of model linearity. These tests have the advantage of accounting for nonlinearity with respect to the predictions. It can be run after the prediction mode has been completed. See Chapter 17.

The **nonlinear-uncertainty mode** calculates nonlinear confidence or prediction intervals. It needs to be the last of a series of runs described in Chapter 17.

The **investigate-objective-function mode** is usually used to investigate difficulties with the regression. It produces output files that list sets of parameter values and associated objective-function values. Graphics that plot the objective function in relation to one, two, or three parameters can be used to investigate problems such as insensitivity, local minima, and extreme parameter correlation.

Files Associated with Running UCODE_2005

As mentioned in Chapter 2, running UCODE_2005 requires three types of input files.

Main input file. In the files provided in the UCODE_2005 distribution (see Appendix C), these are named with the extension '.in'.

Instruction files. Used to read information from the process-model output files. In the examples provided with the UCODE_2005 distribution, these are named with the extension '.instructions'.

Template files. Used to create process-model input files. In the examples provided, these are named with the extension '.tpl'.

In addition, optional input files include:

Data files referenced in the main input file. Data can be in separate files to promote clarity.

Derivatives interface file. Provides instructions for reading some or all sensitivities from a file generated by the process model, rather than having UCODE_2005 calculate sensitivities by perturbation.

fn.xyzt, a file with a location and time for each observation used for plotting results.

Other files provided for by the input blocks of the main input file.

The UCODE_2005 output files are named using the filename prefix defined on the command line. The prefix is represented here using the letters fn. If they exist, output files are replaced without warning when UCODE_2005 is executed. To preserve output files, change the prefix defined on the command line, rename the files, or copy the files into a different directory. The output files include the following.

fn.#uout is the main UCODE_2005 output file for the forward, sensitivity-analysis, and parameter-estimation modes (Table 3). See Table 3 for the main output filename produced for other modes; for all of tem, the filename extension begins with "#".

Data-exchange files. Their suffixes begin with an underscore "_". For example, fn._os and fn._nm are data-exchange files. Chapters 14 and Appendix B provides lists and descriptions of the many data-exchange files produced by UCODE_2005.

Process-model input files generated by UCODE_2005 using template files.

Process-model output files generated from execution of the process model(s).

For the examples distributed with UCODE_2005, the process-model input and output files reside in a separate directory than the UCODE_2005 input and output files.

Calibration and Prediction Conditions

The process model needed to simulate equivalents to observations may be quite different than the process model needed to obtain predictions. The difference may be a different stress, such as different pumpage in a ground-water model. Or it may be a different process, such as when a ground-water model calibrated with head and flow data is used to predict transport. Calibration conditions need to be defined for all modes except the prediction mode. Prediction conditions are needed for five analyses achieved using the modes listed in table 3.

1. **Prediction mode with Sensitivities=no** in the UCODE_Control_Data input file. (with prediction conditions) Use these runs to check calculation of predictions. In the output file, which has filename extension #upred, predictions are compared to reference values.

2. **Prediction mode with Sensitivities=yes** in the UCODE_Control_Data input file, as listed in table 3 (with prediction conditions). This produces scaled sensitivities s that can be used to identify model parameters important to predictions. The scaled sensitivities are printed to data-exchange files described in Chapter 14.

3. **Sensitivity-analysis or parameter-estimation mode** (with calibration conditions), **and prediction mode** (with prediction conditions). Calculate linear confidence and prediction intervals for the predictions using LINEAR_UNCERTAINTY (see Chapter 15).

4. **Advanced-test-model-linearity mode** (with both calibration and prediction conditions). Test model linearity accounting for predictions using CORFAC_PLUS, and MODEL_LINEARITY_ADV (see Chapter 17).

5. **Nonlinear-uncertainty mode** (with both calibration and prediction conditions). Calculate nonlinear confidence and prediction intervals for the predictions.

For 3, two runs of UCODE_2005 are needed: (1) either sensitivity-analysis or parameter-estimation mode, and (2) prediction mode. For both runs, in the Parameter_Groups, Parameter_Data, and Derived_Parameters input blocks identical parameters need to be defined as adjustable and parameters that are not adjusted need to be assigned the same value.

For 4 and 5, both calibration and prediction runs need to be defined. See the instructions for the advanced-test-model-linearity mode in Chapter 17 for additional information.

Typical UCODE_2005 Project Flow

A typical project flow involves the following:

1. Create a forward process model.
2. Create UCODE_2005 files: main input file, instruction files, template files. Possibly a derivatives interface file. (see Chapters 6 through 13)
3. Execute UCODE_2005 in forward mode (Table 3) with Verbose=4 or 5 in the Options input block of the UCODE_2005 main input file. Review the process-model input and output files and fn.#uout to confirm that the substitutions and

extractions are correct. When satisfactory results are obtained, set Verbose=0 unless a problem is encountered.

4. Execute UCODE_2005 in sensitivity analysis mode (Table 3). Review fn.#uout to evaluate sensitivities and correlations. If many sensitivities equal zero, consider the suggestions below in the section 'Troubleshooting'. If some composite scaled sensitivities are less than 1.0 or less than 0.01 times the largest composite scaled sensitivity, or if any correlation coefficients equal 1.00, consider the suggestions in the section "Common ways of improving a poor model" in Chapter 3.

5. Execute UCODE_2005 parameter-estimation mode (Table 3). Review fn.#uout to evaluate the results of the regression as described in Chapter 16. If the regression did not proceed well, consider why, and alter input. See the suggestions in the section "Common ways of improving a poor model" in Chapter 3.

6. Perform residual analysis using RESIDUAL_ANALYSIS as described in Chapter 15 and 16. Evaluate fn.#resan. Often it is useful to include RESIDUAL_ANALYSIS in the batch file with UCODE_2005 in parameter-estimation mode.

7. Simulate predictions using UCODE_2005 in prediction mode (Table 3). Review fn.#upred. Execute LINEAR_UNCERTAINTY using the instructions in Chapter 15. Review fn.#linunc and evaluate the quality of the calibration from the perspective of prediction using the prediction scaled sensitivities.

8. Use the instructions in Chapter 15 to run MODEL_LINEARITY. Look in fn.#umodlin for the modified Beale's measure and an explanation of what it means.

9. See Chapter 17 for how to proceed for the advanced analyses and to calculate nonlinear intervals.

Running RESIDUAL_ANALYSIS

Running RESIDUAL_ANALYSIS requires the following steps

1. Run UCODE_2005 in parameter-estimation mode (see Table 3) and obtain a converged regression. Underscore files need to be produced, so DataExchange=yes is needed by default or designation in the UCODE_Control_Data input block.

2. Run RESIDUAL_ANALYSIS in the same directory using the same filename prefix, fn. The RESIDUAL_ANALYSIS run command is of the form:

path:\ RESIDUAL_ANALYSIS fn

where:

path:\ = the relative or absolute path to RESIDUAL_ANALYSIS.exe on your computer (alternatively you could specify this in your system path variable)

fn = filename prefix used for the UCODE_2005 run (spaces are not allowed in fn, even on operating systems that allow spaces in filenames).

Additional information is provided in Chapter 15.

Running MODEL_LINEARITY

To execute MODEL_LINEARITY, first execute UCODE_2005 twice as indicated below, in the same directory. Underscore files need to be produced, so DataExchange=yes is needed by default or designation in the UCODE_Control_Data input block.

1. Execute UCODE_2005 in parameter-estimation mode (see Table 3) to produce a converged regression.

2. Execute UCODE_2005 in model linearity mode (see Table 3)

3. Execute MODEL_LINEARITY in the same directory using the same filename prefix, fn. The MODEL_LINEARITY run command is of the form:

path:\MODEL_LINEARITY fn

where:

path:\ = the relative or absolute path to MODEL_LINEARITY.exe on your computer (alternatively you could specify this in your system path variable)

fn = filename prefix used for the UCODE_2005 runs (spaces are not allowed in fn, even on operating systems that allow spaces in filenames).

Additional information is provided in Chapter 15.

Running LINEAR_UNCERTAINTY

To execute LINEAR_UNCERTAINTY, first execute UCODE_2005 twice, as indicated below, in the same directory. Underscore files need to be produced, so DataExchange=yes is needed by default or designation in the UCODE_Control_Data input block.

1. Execute UCODE_2005 in parameter-estimation mode (see Table 3) to produce a converged regression.

2. Execute UCODE_2005 in prediction mode (see Table 3).

3. Execute LINEAR_UNCERTAINTY in the same directory using the same filename prefix, fn. The LINEAR_UNCERTAINTY run command is of the form:

path:\ LINEAR_UNCERTAINTY fn

where:

path:\ = the relative or absolute path to LINEAR_UNCERTAINTY.exe on your computer (alternatively you could specify this in your system path variable)

fn = filename prefix used for the UCODE_2005 runs (spaces are not allowed in fn, even on operating systems that allow spaces in filenames).

Additional information is provided in Chapter 15.

Trouble Shooting

Most of the problems encountered when running UCODE_2005 arise from omitting required data items from the input files, listing data in the input files in improper order, specifying incorrect values, or specifying filenames and pathnames incorrectly.

UCODE_2005 is programmed to recognize certain problems related to input data. The problems may be errors that cause execution to stop, or warnings that do not stop execution. The input instructions describe some of these problems. When an error is encountered, an error message is written to the main UCODE_2005 output file and execution stops. Warnings also are printed to the main output file, but execution continues. Some error messages also are printed to the screen and these are often useful in detecting errors in filenames and pathnames.

Other problems commonly encountered result from the regression problem being poorly posed. For example, (1) at least some of the defined parameters are insensitive to the available data, (2) the parameters are extremely correlated, or (3) the model exhibits strong nonlinearity. These problems can be addressed as discussed in Chapter 3 and the references cited therein.

RESIDUAL_ANALYSIS, RESIDUAL_ANALYSIS _ADV, LINEAR_UNCERTAINTY, MODEL_LINEARITY, and MODEL_LINEARITY_ADV use files created by UCODE_2005, so problems with their execution are commonly related to problems in running UCODE_2005. Check to be sure the needed data exchange files have been produced, are located in the appropriate directory, and contain the expected data.

The following sections provide specific directions for detecting and addressing two common problems: incorrect simulated values (either simulated equivalents of observations or predictions) and sensitivities that equal zero when they should not.

What to Do When Simulated Values are Wrong

When simulated values are wrong, they can be checked using two forward mode runs for which keyword Verbose=4 or Verbose=5 in the Options input block (Chapter 6).

First, delete the model input files that are to be replaced using template files. Also delete the model output files from which values are to be extracted and the main UCODDE_2005 output file for the mode being executed (Table 3). Perform the first run and save the newly created files in a different directory.

For the second run, change one or more starting parameter values in the Parameter_Values input block (see Chapter 7) by as much as possible while maintaining a process model that runs correctly (large changes help identify problems). Run UCODE_2005.

Check the process-model input files to make sure they are changed as expected. If not, check the template file. Possible problems are an incorrect template file or that the parameter names listed in the template file do not match the parameter names in the UCODE_2005 main input file. Also check the path to the template file and to the model input file.

Check the process-model output files to see if simulated values are changed. If not, make sure that the input files created by UCODE_2005 are being used by the process model(s).

Finally, check the UCODE_2005 main output file to see if the values extracted by UCODE_2005 are changed as expected. If not, check the instruction file and the path to the model output file specified in the UCODE_2005 main input file.

What to Do When Sensitivities Equal Zero

Another common problem is that the simulated values are correct but sensitivities are unexpectedly equal to zero. One symptom may be that UCODE_2005 fails to complete parameter estimation.

The existence of observation sensitivities equal to zero that should not be zero can be investigated as follows. The data-exchange files listed are created if keyword DataExchange=yes by default or designation in the UCODE_Control_Data input block (Chapter 6). The printing of the sensitivity tables in the main output file is controlled by keywords StartSens, IntermedSens, and FinalSens of the UCODE_Control_Data input block (Chapter 6).

1. It is good practice to use the sensitivity-analysis mode to check for observation sensitivities equal to zero before running the parameter-estimation mode. Check the sensitivity tables printed in the main output file and in data-exchange files with filename extensions _sc, _sd, _s1 and _su. For the sensitivity-analysis mode, sensitivities are calculated using the starting parameter values.
2. If UCODE_2005 fails while estimating parameters and DataExchange=yes by designation or default, the most recent sensitivities are listed in the data-exchange files with filename extensions _sc, _sd, _s1 and _su. Check these files for sensitivities that equal zero when they should not be zero.
3. Sensitivities for all intermediate sets of parameter values can be printed to the main output file by setting IntermedSens=dss in the UCODE_Control_Data input block and executing UCODE_2005 again in the parameter-estimation mode. The sensitivities in the data-exchange files are always from the most recent parameter values.

A value of zero is calculated and printed for the sensitivity by UCODE_2005 if the simulated values read for the perturbed and unperturbed parameters are identical given the number of significant figures printed in the output. If many of the other sensitivities related to this parameter are nonzero, the zero value may simply indicate that the observation is not very important to estimation of the parameter involved, and this is correctly represented by the sensitivity values being equal to zero. In such a situation, no corrective action is needed. If all of the sensitivities for a parameter are zero, a problem may exist. The problem may be related to input file construction problems or something more fundamental.

Input file construction errors can be investigated through the following steps.

a. Make sure that the process model can be run using the command provided in the Mode_Command_Lines input block. Often this is best accomplished by deleting one or more process-model output files to make sure they are being created correctly using the command.

b. Check that input files are being created correctly using the template file. This can be accomplished by deleting the input file before running UCODE_2005 in forward mode and then checking the input file to be sure it has been created with the expected numbers.

c. Make sure that the template file allows enough spaces for the substituted values for the perturbed parameter value to create a change in the input file. If derived parameters are used, this requires consideration of how the parameter change is affected by the equation applied.

d. Check to make sure that the values simulated by the process model are changing and being extracted by UCODE_2005 as intended. Use the procedure described in the previous section of this report.

Five other possible corrective actions are:

(1) Smaller solver convergence criteria can be specified in the application codes to improve the accuracy of the simulated values. This also can increase execution time.

(2) Alter the process-model input files or code so that the values extracted by UCODE_2005 are printed with more significant figures in the process-model output file. Be sure that the values are calculated with sufficient accuracy to make the additional digits meaningful.

(3) The datum of the problem can be changed or normalized, as in Chapter 3 in the section 'Perturbation sensitivities'.

(4) Increase the perturbation amount for the parameter, possibly combined with using more significant digits to represent the parameter in the template files. If large perturbations still result in zero sensitivities, consider the possibility that the parameter has no effect on the observed quantities or that the simulated values are not being properly extracted from the model output. If the latter is suspected, consider the suggestions in the previous section.

(5) Consider the methods for coping with insensitive parameters discussed in Chapter 3 in the section entitled 'Common ways of improving a poor model'.

If available, the first three options are preferable because increasing the perturbation size suggested in the fourth option produces less accurate sensitivities for nonlinear parameters and the fifth option may require a less detailed parameterization.

Chapter 5: OVERVIEW OF UCODE_2005 INPUT INSTRUCTIONS

UCODE_2005 requires a main input file, at least one template file, and at least one instruction file. The template files are used to define parameters to be estimated; the instruction files describe how to extract values from process-model output files. The main input file can read data from other files to facilitate data management. Additionally, a Derivatives Interface input file is needed if any sensitivities are read from files produced by a process model. For example, MODFLOW-2000 with its Sensitivity Process can be used to calculate accurate sensitivity-equation sensitivities.

This chapter begins by describing the main UCODE_2005 input file. First basic input concepts are explained. Then a detailed discussion of input for the Main input file is provided.

Main Input File

As discussed in Chapter 4, the main input file for UCODE_2005 is named on the command line, which is of the form:

C:\wrdapp\ucode_2005_1.000\bin\UCODE_2005.exe   ex1.in   ex1

The main input file can contain comments that the program will ignore. Comment lines begin with # in the first column. No spaces or other characters can precede the # on comment lines. Comment lines can be inserted almost anywhere in the main input file.

Input blocks

The main input file includes input blocks with the basic structure:

```
Begin blocklabel [blockformat]
  Blockbody: many lines OR, when blockformat is 'files', a list
  of one or more files
End blocklabel
```

The brackets around blockformat indicate that this variable is optional. Square brackets are used to identify optional variables throughout this document. All keywords are case-insensitive on input and space-delimited.

The variables blocklabel and blockformat are defined in the following sections. The definition of blockbody depends on blocklabel; the possibilities are described in the following chapters.

The input blocks described in this report are part of the JUPITER API or are designed using the conventions established as part of the API.

Blocklabel

The variable blocklabel identifies the purpose of the data block and the data it can contain. This chapter provides general information about blocklabel. The data needed for each blocklabel is described in subsequent chapters.

Some blocklabels are required, as indicated in table 5. Additional capabilities of UCODE_2005 can be accessed by using additional blocklabels.

If a blocklabel is misspelled, the data are ignored and defaults assigned. Ignoring unneeded input blocks allows great flexibility for the sequences of runs common with UCODE_2005 because most input blocks do not need to be removed even if they are not needed in a subsequent step. More generally, this feature allows different applications of the JUPITER API to use the same or very similar input files. The drawback is that an input block is ignored if the blocklabel is misspelled. To check, review the echo of the input printed in the UCODE_2005 main output file.

Blockformat

The variable blockformat defines the structure of the data presented. The options are listed in Table 5. The default blockformat is Keywords, but it is urged that the blockformat be listed specifically to reduce confusion.

The input blocks used in UCODE_2005 are very flexible. One resulting difficulty is that if the blockformat specified does not match the format used, the information in the data block is ignored and generally no error message is printed. For example, if blockformat 'Keyword' is specified by default or designation, data organized in blockformat 'Table' is ignored. The problem can be detected by inspecting the echo of the input in the main UCODE_2005 output file.

Table 4. Blocklabels of the main input file for UCODE_2005.
[Bold type and grey shading identify required input blocks, as qualified by footnotes 4 and 5; the other input blocks are optional.]

Chapter[1]	Purpose	Blocklabel	Default column order[2]
[4]6	Define UCODE_2005 operation	Options	No
		Merge_Files[3]	No
		UCODE_Control_Data[3]	No
		Reg_GN_Controls[3]	No
		Reg_GN_NonLinInt[3,4]	No
		Model_Command_Lines	No
7	Define parameters	Parameter_Groups[3]	No
		Parameter_Data[3]	Yes
		Parameter_Values[3]	Yes
		Derived_Parameters[3]	Yes
8	Define observations	Observation_Groups	No
		Observation_Data[5]	Yes
		Derived_Observations	Yes
	Define predictions	Prediction_Groups	No
		Prediction_Data[6]	Yes
		Derived_Predictions	Yes
9	Define prior information	Prior_Information_Groups	No
		Linear_Prior_Information	Yes
10	Define variance-covariance matrices to weight groups of observations or prior information with correlated errors.	Matrix_Files	No
11	Interact with process-model input and output files.	**Model_Input_Files**	Yes
		Model_Output_Files	Yes
12	Run process model(s) using multiple processors	Parallel_Control	No
		Parallel_Runners	Yes

[1] Chapter that describes these input blocks.

[2] 'Yes': the input block has a default column order. With blockformat TABLE, these blocks can contain data without column labels for selected keywords if the data are in default order. Keywords defined in the JUPITER API are supported. Keywords added for UCODE_2005 always need column labels. 'No': the input block has no default column order, use column labels.

[3] Programmers: These are UCODE_2005 input blocks, not part of the JUPITER API.

[4] Reg_GN_NonLinInt is documented in Chapter 17.

[5] Required for all UCODE_2005 modes except the prediction mode, for which all observation input blocks need to be omitted. The prediction mode is defined when prediction=yes in the UCODE_Control_Data input block. For mode definition, see table 3 and Chapter 17.

[6] Required for UCODE_2005 modes prediction (table 3) and advanced-test-model-linearity and nonlinear-uncertainty (Chapter 17). For other modes, all prediction input blocks need to be omitted. The modes are defined with keywords in the UCODE_Control_Data input block. For mode definition, see table 3 and Chapter 17.

Table 5. Blockformat options.

Blockformat	Prescribed input format
KEYWORDS	Blockbody consists of a series of lines of the form: Keyword=value Under some circumstances there are restrictions on how the lines are ordered; see the input block instructions. If no blockformat is specified, KEYWORDS is assumed, but it is advisable to explicitly identify the block format to reduce errors. Comments are allowed.[1,2]
TABLE	Blockbody consists of a table of data that may have labels on the columns and may be read from the main input file or from another input file. See the text for additional information. Comments are allowed right after the BEGIN statement but not in the rest of the input block. [1]
FILES	Blockbody consists of the pathname for one or more files. Comments are allowed.[1,2] To allow the format to be specified, the contents of each of the listed files needs to begin with a 'Begin Blocklabel [Blockformat]' line and end with an 'End Blocklabel' line. The Blocklabel needs to be the same as in the 'Begin Blocklabel FILES' block within which the files are listed. See the section "Observation_Data Input Block" for an example.

[1] Comments are separate lines starting with a # in the first column. No blank lines are allowed within any input blocks.

[2] Comments can be inserted anywhere within the input block.

Blockbody

blockbody contains data or the names of files from which the data are to be read. The format of the data is determined by *blockformat*.

The meaning of the data provided is defined using keywords. Keywords that are not recognized are ignored. This allows a constructed input block to be used for multiple purposes without modification. It also means that misspelled keywords are not flagged as errors and default values will be used if keywords are misspelled. This problem can be identified by reviewing the echo of the input file in the main UCODE_2005 output file. For many keywords, a default is available and is used if the keyword is omitted.

Blockformat KEYWORDS

If blockformat is specified as **KEYWORDS**, *blockbody* is expected to be a series of phrases of the form *keyword=value*. For example, PARAMNAME=K1. There can be spaces on each side of the equal sign. Phrases can occur on separate lines or can occur on the same line if they are separated by spaces.

Some keywords can appear in any order while other keywords indicate the need for associated data to be provided either through a subsequent set of keywords or by other means. The options available depend on the input block, as described in the following chapters.

An example of a keyword that indicates the need for associated data occurs for *blocklabel* **Parameter_Data**. Each time the keyword **PARAMNAME** appears, a parameter is defined and a related set of data is needed. For a parameter, the data can be defined by keywords that follow keyword **PARAMNAME** in the Parameter_Data input block, or by data provided in the Parameter_Groups input block. The keyword **PARAMNAME** and associated data are repeated for each parameter. This can be tedious, and blockformat TABLE is often more convenient in this circumstance.

Here is a simple example input block using blockformat keywords. The keywords are defined in Chapter 6 in the section in the Options input block input instructions.

```
BEGIN Options Keywords
Verbose=0
Derivatives Interface = "tc1.derint"
END Options
```

Blockformat TABLE

If *blockformat* is specified as **TABLE**, the first non-comment line of *blockbody* is in the format:

```
NROW=nr NCOL=nc [COLUMNLABELS] [DATAFILES=nfiles]
[GROUPNAME=gpname]
```

The format of the rest of the blockbody depends on whether DATAFILES is listed, as shown in Table 6.

Table 6. For blockformat *TABLE*, the format of blockbody after the first line without and with the optional keyword DATAFILES.

Without DATAFILES keyword	With DATAFILES keyword
[column-name] [column-name]...	[column-name] [column-name]...
val val ...	pathname [SKIP=nskip]
val val ...	pathname [SKIP=nskip]
... ...	...
number of lines: nr	number of lines: nfiles

Definition of keywords and variables:

NROW and **NCOL** are required keywords.

nr is the number of rows in the table.

nc is the number of columns in the table.

COLUMNLABELS is an optional keyword.

> **COLUMNLABELS** omitted: A default column order is used to identify the data in the columns of the table. Default column orders are only available for the *blocklabel*s identified in section 'Blocklabels'. If a default column order is not available (Table 4, last column) COLUMNLABELS is required.

> **COLUMNLABELS** listed: Column names are used to identify the data in the columns of the table. Data is read for columns with column names that are equivalent to keywords for this *blocklabel*. The keywords for each input block are defined in the following chapters. Data in columns with other labels are ignored. This allows data sets to contain columns that are not used by UCODE_2005. However, it also means that misspelled keywords are not flagged as errors and default values will be used if keywords are misspelled.

DATAFILES is an optional keyword.

> **DATAFILES** omitted: *nr* rows of data are read as shown in column 1 of Table 6. Each *val* is a data value. The data type expected for *val* depends on the *blocklabel* and possibly on *column-name*. All data values for a row need to be on one line of the file. One line can contain up to 2,000 characters.

> **DATAFILES** listed: A list of `file pathnames` is read, as shown in the second column of Table 6. The number of pathnames read equals *nfiles,* for example, **DATAFILES**=2. Each *pathname* is the path to a file from which rows of data are read. Paths with spaces need to be enclosed in double quotes. Each file needs to contain rows of data in columns in either the default column order or the order defined by the *column-name* entries, if specified. Data read from all files are combined as if read from one file. Each file is read in order until *nr* rows of data have been read. If **SKIP=***nskip* is specified, *nskip* lines at the beginning of the file are ignored, and reading of data starts on the following line.

GROUPNAME is an optional keyword

> For blocks that use groups, **GROUPNAME=***gpname* can be used to assign a group name to all rows in the table. *gpname* is the group name. If **GROUPNAME=***gpname*

is present, **GROUPNAME** will not be in the default list of columns and can not be included with the **COLUMNLABELS** option.

Here is a simple example input block using blockformat table. The keywords are defined in Chapter 7.

```
BEGIN Parameter Values TABLE
# These values override values in Parameter Data input block
  nrow=9  ncol=2  columnlabels
  paramname   startvalue
  Wells TR    -1.1000
  RCH Zone 1  6.3072E+1
  RCH Zone 2  3.1536E+1
  Rivers      1.2000E-3
  SS 1        1.3000E-3
  HK 1        3.0000E-4
  Vert K CB   1.0000E-7
  SS 2        2.0000E-4
  HK 2        4.0000E-5
END Parameter Values
```

blockformat **FILES**

If **blockformat** is specified as **FILES**, the input block can contain one or more lines, each containing a pathname to a file. Lines with # as the first character are interpreted as comments and are ignored. Data read from all files in the list are combined to create one blockbody. The data need to be composed of blocks with Begin and End statements.

Data can be read from files in two ways. The mechanisms and their characteristics are described in Table 7.

Table 7. Alternatives for reading data from files.

Blockformat table With **DATAFILES**	Blockformat files
There is only one Begin blockformat and End blockformat block.	There can be more than one Begin blockformat and End blockformat block.
All data are read as a table.	Blockformat can change based on the designations in the Begin statements

Here is a simple example input block using blockformat files. The keywords are defined in Chapter 8 in the Observation Data input block instructions.

```
BEGIN OBSERVATION DATA FILES
tc1.hed
tc1.flo
END OBSERVATION DATA
```

Files tc1.hed and tc1.flo are read. For example, file tc1.flo might be as follows.

```
BEGIN OBSERVATION DATA TABLE
  NROW=3  NCOL=4  COLUMNLABELS
  Obsname      obsvalue  statistic      equation
  flow.ss      -4.4      0.4
  flow.t3      -4.1      0.38
  flow.t12     -2.2      0.21
  flow.t3 ss    0.3      0.55          flow.t3 - flow.ss
  flow.t12 ss   2.2      0.45          flow.t12 - flow.ss
END OBSERVATION DATA
```

Additional Input Files

The additional input files listed in Table 8 also may be needed. They are described briefly below and in detail in subsequent chapters. Some of the filenames are determined completely by the users, others (here, fn.omit and fn.xyzt) are prescribed. By convention, no underscore or # signs are used in UCODE_2005 prescribed file extensions of user-produced files.

The derivatives interface file allows sensitivities calculated by a process model to be used in the UCODE_2005 calculations. This can be advantageous if the process model can calculate sensitivities using a more accurate or computationally efficient method than the perturbation methods available in UCODE_2005. For example, the Sensitivity Process of MODFLOW-2000 can be used to calculate sensitivity-equation sensitivities that are generally more accurate and in some circumstances can be calculated more efficiently than perturbation sensitivities.

Template files are used to interact with process-model input file(s), as discussed in Chapter 11.

The fn.xyzt file provides the user the opportunity of defining locations and times for each observation. The fn is the prefix name defined on the command line; no other name is allowed for this file. The location and time of observations are not needed for UCODE_2005 to perform its calculations. However, evaluation of model fit generally requires that quantities such as residuals and weighted residuals be plotted on maps or against time. To assist the user in performing these essential analyses, UCODE_2005 looks for a file named fn.xyzt in the directory in which UCODE_2005 is executed. If found, the file is read, residuals and weighted residuals are associated with each listed observation, and output file fn._xyztwr is produced. The contents of the fn.xyzt file do not in any way affect the calculations performed by UCODE_2005.

In many process models, simulated values are assigned specific numbers when the value can not be simulated. This may occur, for example, in a ground-water model for what is expected to be a saturated-zone hydraulic head from a part of the model that becomes desaturated. In such a situation, any related observation generally can not be used in regression. The input file fn.omit lists numbers that indicate that the process model can not simulate a value. When a simulated equivalent equals one of these numbers,

UCODE_2005 omits the observation from the regression and a message is printed in its place in the UCODE output file. For fn.omit, the fn is the prefix name defined on the command line; no other name is allowed for this file.

Most input blocks allow data to be read from other files. The flexibility of the input blocks means that many such files could be defined.

Table 8. Additional input files, their purpose, the label of the input block that uses the file(s), and the chapter that provides detailed input instructions.
[fn.xyzt and fn.omit, filenames for which the fn needs to be the same as the filename prefix specified on the command line; the filename extensions need to be as indicated. By convention, no underscore or # signs are used in the filename extensions of input files.]

File name or description[1]	Purpose	Blocklabel (Chapter)
Derivatives Interface File (optional)	Instructions for reading derivatives calculated by a process model.	Options (6)
Template files	Used to create process-model input files that reflect starting parameter values defined in the Parameter_Group, Parameter_Data input block, or as updated by regression.	Model_Input_Files (11)
fn.xyzt (optional)	Provide spatial and temporal coordinates for observations and predictions.	No input block is involved. UCODE_2005 searches for a file called fn.xyzt in the directory from which it is executed. (13)
fn.omit	A list of real numbers. If simulated equivalents equal any of these values, the associated observation is omitted from the regression.	Reg_GN_Controls (6)
Other files (optional)	To provide data for input blocks using: (1) blockformat table with the datafiles keyword or (2) blockformat files. The path of the file is defined in the input block.	Most input blocks. (6-12)

[1] In the example of Appendix C, derivative-interface files have filename extension .derint and template files have extension .tpl.

Chapter 6: INPUT TO CONTROL UCODE_2005 OPERATION

UCODE_2005 can be used to perform a number of tasks related to sensitivity analysis, parameter estimation, and uncertainty analysis. As many as four input blocks can be used to define what is accomplished in a given execution.

The Options input block can be used to control what is written to the main output file and to identify a derivatives interface file, which provides directions for reading sensitivities produced by the process model.

The UCODE_Control_Data input block controls most of what can be accomplished with UCODE_2005.

The Reg_GN_Controls input block controls the modified Gauss-Newton regression, including its double-dogleg option.

The Model_Command_Lines input block defines how to run the process model(s) and is the only input block of this chapter that is always required.

Options Input Block: Control Main Output File and Read Sensitivities (optional)

The Options input block can be used to control the following:

1. The information written to the main output file

2. Sensitivities read for all or some parameters from a process-model output file that calculates sensitivities. For example, MODFLOW-2000 calculates sensitivities using the sensitivity-equation method which is generally more accurate than the perturbation methods of UCODE_2005.

The three keywords are usually read using the KEYWORDS blockformat.

Verbose   - Flag that controls what is written to the UCODE_2005 main output file as follows. The default is Verbose=3 to provide information for new applications and users, but Verbose=0 is suggested for most circumstances.

Verbose	Output
0	No extraneous output.
1	Warnings.
2	Warnings, notes.
3 (default)	Warnings, notes, echo selected input.
4	Warnings, notes, echo all input. Includes all values read from process-model output files.
5	Warnings, notes, echo all input, plus some miscellaneous information. Includes all values read from process-model output files.

The paths defined using the following two keywords can be up to 2,000 characters long and need to be surrounded by double quotes if there are any spaces. The paths can be relative or absolute.

Derivatives_Interface - Filename or path to derivatives-interface input file. Construction of the derivatives interface input file is described in Chapter 13. No default. If keyword Derivatives_Interface is not included in the Options input block, no Derivatives Interface input file is read. The Derivatives_Interface keyword needs to be omitted if SenMethod=1 or 2 for all parameters in the Parameter_Groups and Parameter_Data input blocks (see Chapter 7).

PathToMergedFile - Filename or path of a file into which data are merged from the files listed in the following Merge_Files input block. If the file exists, it is replaced.

The Derivatives Interface input file described in Chapter 13 only allows one file to be read. Sensitivities located in more than one file can be read by (1) using

PathToMergedFile to define a file, (2) listing the files with sensitivities in the Merge_Files input block described next, and (3) in item 1 of the Derivatives Interface input file described in table 13 of Chapter 13, listing the path specified by PathToMergedFiles.

Example of an Options input block:

```
BEGIN Options Keywords
Verbose=0
Derivatives_Interface = "tc1.derint"
END Options
```

In this example, the derivatives interface file is located in the directory in which UCODE_2005 is executed so no path is specified.

Merge_Files Input Block (Optional)

The Merge_Files input block provides the information needed to combine data from a set of files into one file. In UCODE_2005, a merged file is needed because the Derivatives Interface input file can read sensitivities from only one file, and it is common to need sensitivities that are located in more than one file. See the instructions for keyword PathToMergedFile in the Options input block for additional information.

The data in the files to be combined in the Merge_Files input block need to be organized such that they can all be read with the same DerFormat specified in item 5 of the Derivative Interface input file (see table 13, Chapter 13).

The Merge_Files input block has two keywords.

PathToFile — Path of a file. The path can be up to 2,000 characters long and needs to be surrounded by double quotes if it contains any spaces. The path can be relative or absolute.

SkipLines — Number of lines that are to be omitted from the top of the file before it is appended onto the end in the combined file. Usually these are header lines. Default=0.

If SkipLines=0 for the first "PathToFile", the header lines in the combined file are the header lines from the first file listed. For subsequent files listed, any header lines need to be omitted; only lines of data from these files can be added to the merged file.

If Blockformat KEYWORDS is selected by designation or default, keywords defining a file in the Merge_Files input block need to be grouped together. The PathToFile keyword needs to be the first keyword on a new line. PathToFile and SkipLines, if needed, are repeated to define multiple parameters. SkipLines is needed only to change the default value.

If blockformat TABLE is selected without indicating ColumnLabels, the default column order is the order in which the keywords are defined above and listed below. No columns are ignored and a column for each keyword is needed. If ColumnLabels are indicated, the column labels can appear in any order; the ParamName keyword need not be first, though it often is first.

Default column order: `PathToFile SkipLines`

Example of a Merge_Files input block:

```
BEGIN Merge Files TABLE
NROW=2 NCOL=2 COLUMNLABELS
PATHtoFILE                SKIPLINES
../../CALIBRATION/EX1. SU      0
../../PREDICTION/EX1. SU       0
END Options
```

UCODE_Control_Data Input Block: Define the Task and Output (optional)

The UCODE_Control_Data input block defines the operations pursued by UCODE_2005 and defines some labeling for data-exchange files.

Variables can be read in any order. Included variables generally are read using the default keyword format. TABLE format requires COLUMNLABELS because there is no default order for these variables.

Only variables for which the default values are not acceptable need to be included.

Chapter 17 describes two additional keywords needed to pursue advanced testing of model linearity and nonlinear confidence intervals.

Keywords are not case sensitive.

ModelName   - Identifies the model. Up to 12 characters. Default=generic.

The following three keywords are used to define the units used in the model in the main output file and the _dm data-exchange input file. Each can be up to 12 characters long. The units defined are used as labels; UCODE_2005 performs no unit conversions. Defining units here does not affect the solution. The default is NA, indicating that the units are undefined. Specifying units can reduce confusion and modeling mistakes.

ModelLengthUnits - Defines the length unit. For example, ft for feet, m for meters, km for kilometers, and mi for miles.

ModelMassUnits - Defines the mass unit. For example, mg for milligrams, g for grams, kg for kilograms, lb for pounds.

ModelTimeUnits - Defines the time unit. For example, s for seconds, min for minutes, d for days, mo for months, and yr for years.

If the following keywords, Sensitivities, Optimize, Linearity, LinearityAdv, Prediction, NonlinearIntervals, and SOSSurface, are all "no" by designation or default, the forward mode is executed. For model calibration, the forward mode is typically performed first to check the execution of the process model(s) from the command lines, substitution of parameter values, and reading of model output values. See Table 3 for the UCODE_2005 modes produced when these keywords equal "yes".

Sensitivities - yes: calculate sensitivities or read them from a derivatives interface file listed in the Options input block; UCODE_2005 main output file is fn.#uout. no: do not calculate sensitivities. Default=no.

Optimize - yes: estimate parameters; UCODE_2005 main output file is fn.#uout. no: do not estimate parameters. Default=no.

Linearity   - yes: execute the test-model-linearity mode and produce the data-exchange file fn._b2 needed by the program MODEL_LINEARITY (Chapter 15). UCODE_2005 main output file is fn.#umodlin. no: do not produce the file. Default=no.

The test-model-linearity mode needs to be run in a directory containing data-exchange files from a successful parameter-estimation mode run. Designate the same filename prefix on the command line that was used for the other runs.

Prediction   - yes: determine predicted values and their sensitivities using the Prediction input blocks of Chapter 8; UCODE_2005 main output file is fn.#upred. The output is often used by postprocessor LINEAR_UNCERTAINTY. *This may require use of alternative process-model input files and associated template files.* Default=no.

LinearityAdv   - 'conf' or 'pred': execute the advanced-model-linearity mode and produce files needed by the program MODEL_LINEARITY_ADV (Chapter 17). UCODE_2005 main output file is fn.#umodlinadv_*, where * is 'conf' or 'pred'. Default=no.

Run the advanced-test-model-linearity mode in a directory containing data-exchange files from successful parameter-estimation and prediction mode runs as well as a CORFAC_PLUS run (Chapter 17). Designate the same filename prefix on the command line that was used for the other runs. Use LinearityAdv=conf to use parameter values from a data-exchange file with extension _b1advconf. Use LinearityAdv=pred to use parameter values from a data-exchange file with extension _b1advpred. The _b1adv* files are produced by CORFAC_PLUS. See Chapter 17 for additional information.

NonlinearIntervals - yes: calculate nonlinear confidence intervals using the instructions in Chapter 17. no: do not calculate nonlinear intervals. NonlinearIntervals=yes results in a full regression for each interval limit. Default=no.

SOSSurface   - yes: calculate sum-of-squared weighted residuals objective function values for the sets of parameter values defined as in (a) below. file: calculate sum-of-squared-weighted-residuals objective-function values for the sets of parameter values defined as in (b) below. no: do not perform these calculations. Default=no.

When SOSsurface =yes or file, the UCODE_2005 main output file is fn.#usos and computed values are printed to fn._sos. For each set of parameter values, a forward command is executed. A command with Purpose=forward needs to be defined in the Model_Command_Lines input block. Use these options to explore the objective function. Commonly this is accomplished using contour maps for two parameters at a time.

(a) The sets of parameter values are controlled by the keywords Adjustable, SOSIncrement, LowerConstraint, and UpperConstraint defined in the PARAMETER_GROUPS or PARAMETER_DATA input block, as described in the PARAMETER_DATA input block instructions. The number of sets (and the number of runs) equals the product of SOSIncrement for each parameter with Adjustable=yes, and can become large.

(b) The sets of parameter values are read from SOSFile. The parameters listed need to have adjustable=yes in the PARAMETER_DATA input block; not all adjustable parameters need be listed. If a listed parameter is not adjustable an error message is printed and execution stops.

SOSFile - Filename or absolute or relative pathname of a user-created file needed when SOSsurface=file. Up to 2000 characters; case sensitivity depends on the operating system.

File SOSFile needs to contain a line with the number of parameters (equal to the number of columns in the file), a line with the parameter names, and lines that each contains a set of parameter values -- one parameter value for each listed parameter name. The number of sets of parameter values is determined by the number of lines in the file; reading continues to the end of the file. For example, an SOSFile with one set of parameter values might contain:

```
5
par1  par2  par3  par4  par5
1.394 4.932 9.664  1546.987  0.3496E-06
```

StdErrOne - yes: calculate statistics without using the value of the calculated standard error. Instead, use a value of 1.0. no: calculate statistics using the standard error calculated from the listed observations. This is typically only 'yes' for projects designed to explore the type, location, and timing of data that may be most useful for estimating parameters. Default=no.

EigenValues - yes: calculate and print eigenvalues and their associated eigenvectors to the main output file; no: do not do this calculation. Default=yes.

The following three keywords control printing of tables of observations, simulated values, and residuals to the main output file before, between, and after parameter-estimation iterations.

StartRes - Controls printing for the starting parameter values. Default=yes.

IntermedRes - Controls printing after each parameter-estimation iteration. Default=no.

FinalRes - Controls printing for the final parameter values. Default=yes.

The following three keywords control printing of sensitivity tables to the main output file before, between, and after parameter-estimation iterations. All three keywords have the same options:

Option	Table(s) of sensitivities printed in the UCODE_2005 main output file
css	Composite scaled sensitivities
dss	Dimensionless and composite scaled sensitivities
onepercentss	One-percent scaled sensitivities and composite scaled sensitivities
allss	Composite, dimensionless, and one-percent scaled sensitivities
unscaled	Unscaled sensitivities and composite scaled sensitivities
all	All tables listed above
none	Nothing

StartSens   - Controls printing for the starting parameter values. Default=dss.

IntermedSens   - Controls printing after each parameter-estimation iteration. Default=none.

> Use IntermedSens=yes to ensure that sensitivities are available from the most recent parameter-estimation iteration if the process model fails during a parameter-estimation iteration. To control sensitivity calculations when regression does not converge, use keyword Stats_On_Nonconverge of the Reg_GN_Controls input block.

FinalSens   - Controls printing for the final parameter values. Default=dss.

The creation of most data-exchange files is controlled by the mode executed (see table 3) and keyword DataExchange. For the sensitivity-analysis mode, a different set of files is produced if CreateInitFiles=yes regardless of the designation f DataExchange.

DataExchange   - yes: generate the data-exchange files containing data for graphical and numerical analysis. no: do not produce the files. Default=yes.

CreateInitFiles   - Applies only for sensitivity-analysis mode (table 3). yes: generate data-exchange files with filename extensions _init, _init._mv, _init._su, and, if prior information is defined, _init._supri. These files are used by runs described in Chapter 17. No other data-exchange files are produced. no: do not generate these files; generate other data-exchange files if keyword DataExchange=yes. Default=no.

Example of a UCODE Control Data input block:

```
BEGIN UCODE CONTROL DATA KEYWORDS
  ModelName=ex1fullprior
#Units
  ModelLengthUnits=m
  ModelMassUnits=na
  ModelTimeUnits=d
#Performance
  sensitivities=yes
  optimize=yes
  StdErrOne=no
  EigenValues=yes
#Printing and output files
  StartRes = yes
  STARTSens = css
  IntermedRes=no
  IntermedSens= css
  FinalRes= no
  FINALSens= css
  DataExchange=yes
END UCODE CONTROL DATA
```

Reg_GN_Controls Input Block: Control Parameter Estimation (optional)

The Reg_GN_Controls input block controls the performance of the modified Gauss-Newton regression method of estimating parameter values for the UCODE_2005 parameter-estimation mode.

The variables can be read in any order. Included variables generally are read using the default keyword format. TABLE format requires COLUMNLABELS because there is no default order for these variables.

Only keywords for which the default values are not acceptable need to be included. Keywords are not case sensitive.

Three keywords are used to control when parameter-estimation iterations stop. Ideally, convergence is achieved by satisfying the TolPar criterion and TolSOSC equals 0.0. However, values of 0.01 to 0.1 for TOLSOSC can be useful in the early stages of model calibration to stop parameter-estimation iterations when they are not improving model fit.

TolPar — Tolerance based on parameter values: parameter-estimation iterations stop if the maximum fractional change in parameter values between parameter-estimation iterations is less than the value of TolPar. Default=10^{-2}.

The fractional change is defined relative to the native parameter value at the beginning of the iteration. A value of 0.01 requires the fractional change for all parameters to be less than 1 percent. The value specified here applies to all parameters for which TolPar is not defined in the Parameter_Data input block. Use the Parameter_Data input block to define unique values for selected parameters. To obtain n significant digits in the estimated parameters, set TolPar=10^{-n}. If the parameter value equals 0.0, a value of 1.0 is used.

TolSOSC — Tolerance based on changes to model fit: parameter-estimation iterations stop if the fractional decline in the sum-of-squared weighted residuals over three parameter-estimation iterations is less than TolSOSC. A value of 0.01 requires the reduction to be less than 1 percent over three parameter-estimation iterations. If TolSOSC=0.0, it is not used. Default=0.0.

MaxIter — Maximum number of parameter-estimation iterations allowed before stopping. Default=5.

Two keywords restrict how much parameter values can change in one parameter-estimation iteration.

MaxChange — Maximum fractional amount parameter values are allowed to change between parameter-estimation iterations. The value specified here applies to all parameters; use the Parameter_Data input block to define a unique MaxChange for each parameter. Default=2.0, which means that parameter values can change as much as 200 percent.

MaxChangeRealm - Native, MaxChange applies in native space. Regression: MaxChange applies in regression space. In regression space MaxChange applies to log-transformed values for log-transformed parameters. Default=Native.

Three keywords are used to calculate the Marquardt parameter, which is used to improve ill-posed regression problems (Cooley and Naff, 1990; Hill and Tiedeman, in press, Chapter 5; Hill, 1998. p. 8). The Marquardt parameter, μ, is set to zero at the beginning of each parameter-estimation iteration. If the downgradient direction on the sum-of-squared-residuals surface and the parameter update vector are greater than MqrtDirection using the current value of μ, then μ is increased as:

$$\mu^{new} = \text{MqrtFactor } \mu^{old} + \text{MqrtIncrement} \qquad (8)$$

The Marquardt parameter can be increased multiple times within a parameter-estimation iteration.

MqrtDirection - Angle (in degrees) between downgradient direction on the sum-of-squared-residuals surface and the parameter update vector above which the Marquardt parameter is applied. If this angle approaches $90°$, regression is unlikely to make progress. Altering the search direction with the Marquardt parameter often improves the situation. Default=$85.4°$.

MqrtFactor - See equation 8 for the Marquardt parameter. Default=1.5.

MqrtIncrement - See equation 8 for the Marquardt parameter. Default=0.001.

The following three keywords control the quasi-Newton updating described by Hill and Tiedeman (in press, Appendix B) and Hill (1998, Appendix B). Quasi-Newton updating occasionally produces convergence for difficult problems, though more often greater improvement can be achieved using the double-dogleg method activated by keywords listed later in this section or the suggestions described in the section "Common Ways of Improving a poor model" of Chapter 2.

QuasiNewton - yes: use quasi-Newton updating as indicated by the criteria below. no: never use quasi-Newton updating. Default=no.

The following two keywords are used only if QuasiNewton=yes. If either of these two criteria is met for a parameter-estimation iteration, Quasi-Newton updating is used for that and all subsequent iterations.

QNiter   - Number of parameter-estimation iterations executed before including Quasi-Newton updating. A non-zero value is needed if QuasiNewton=yes. Default=5.

QNsosr   - Fractional change in the sum-of-squared weighted residuals over two parameter-estimation iterations below which Quasi-Newton updating is used. A value of 0.01 indicates that Quasi-Newton updating occurs if the sum-of-squared weighted residuals changes less than 1 percent over two iterations. Default=0.01.

OmitDefault   - The number of values to read from user-created file fn.omit. When a simulated equivalent equals one of these numbers, the observation is omitted from the regression and a message is printed in its place in the UCODE output file. This is useful if the application code prints a default value for items that cannot be calculated. Default=0.

The user-created file needs to have the name fn.omit, where fn is replaced by the filename prefix specified on the UCODE_2005 command line. It needs to be located in the directory where UCODE_2005 is executed. The file needs to contain OmitDefault real numbers and be constructed as follows. The first line is BEGIN OMIT_DATA. Starting with the next line, enter one value per line, with no blank lines. End with the line END OMIT_DATA.

For example, if OMITDEFAULT=2, the file fn.omit may contain:

```
#
BEGIN OMIT DATA
1.D30
999.
END OMIT DATA
```

Stats_On_Nonconverge – yes: when parameter estimation does not converge in the maximum number of iterations, calculate final sensitivities and calculate and print final statistics. For perturbation sensitivities, central differences are used so that the process model is executed $(2 \times NP)+1$ more times, where NP is the number of estimated parameters. no: when parameter estimation does not converge do not calculate and print final statistics. Default=yes.

Three keywords control dynamic omission of insensitive parameters from the regression. If some adjustable parameters are insensitive, parameter estimation is likely to perform poorly without dynamic omission of insensitive parameters.

OmitInsensitive – yes: use composite scaled sensitivities (CSS) to omit parameters from the regression and reinclude them. no: always include all adjustable parameters. Default=no.

Omit parameter j if $CSS_j < (\text{MinimumSensRatio} \times CSS_{max})$, where CSS_j is the CSS for parameter j, CSS_{max} is the largest CSS for any parameter. Affected parameters are held at the last estimated value. Parameter j is included again if ReincludeSensRatio>0.0 and $CSS_j > (\text{ReincludeSensRatio} \times CSS_{max})$. Sensitivities for parameter j are calculated for each iteration, even when it is held constant.

MinimumSensRatio – Used as described for OmitInsensitive. Default=0.005.

ReincludeSensRatio – Used as described for OmitInsensitive. Default=0.02.

If any observation is assigned a WtOSConstant>0 in the Observation_Data input block, its weight is calculated using simulated values and the following keyword is used.

TolParWtOS – TolParWtOS×TolPar equals the parameter-change threshold below which simulated values are used to calculate weights on observations with WtOSConstant>0. Above this threshold, observed values are used to calculate the weights. Default=10.

Three keywords control the trust-region modification of Gauss-Newton regression with the step size determined by the double-dogleg strategy. The trust region method and a number of step size strategies are described by Dennis and Schnabel (1996). Mehl and Hill (2002) show that the trust-region method with the double dogleg strategy can decrease the number of iterations needed by a factor of two and can produce solutions in difficult problems relative to modified Gauss-Newton without these methods. For difficult problems, the suggestions in the section "Common ways of improving a poor model" of Chapter 3 of this report may also be useful. If TrustRegion=no, then MaxStep and ConsecMax are ignored.

TrustRegion – Dogleg: use the trust-region modification with the double dogleg step-size strategy. no: do not use this modification. Default=no.

When TrustRegion=Dogleg, the following apply. Parameters can be log-transformed. Dynamic omission of insensitive parameters and quasi-Newton updating can be used. The convergence criteria are TolPar and TolSOSC, used as described above. There is one restriction: Parameter values can not be constrained; defined constraints are ignored. The update method used for each parameter-estimation iteration is listed in the output file. If the method is always "Full Newton", shorter execution time can be achieved with TrustRegion=no.

MaxStep – Maximum allowable step size. For the double-dogleg method, the default is a function of the parameter values, and is printed in the

UCODE_2005 main output file. The default is used if the MaxStep keword is omitted or if it is assigned a negative value; for example, MaxStep=-1. If the regression is moving too slowly, assign MaxStep a value that is larger than the default. If the regression is too erratic, assign MaxStep a value that is smaller than the default.

ConsecMax - Maximum number of times that MaxStep is used consecutively before execution stops. Default=5.

Use of the maximum allowable step size (MaxStep) by the regression is taken to indicate that either (1) the problem is very poorly defined so that extreme and probably unrealistic parameter values are being estimated or (2) MaxStep is too small and impeding progress of the regression. Instead of incurring many regression iterations, stopping the regression provides the user the opportunity to (1) reevaluate the problem or (2) increase MaxStep, respectively.

Example of a Reg_GN_Controls input block:

```
BEGIN REG GN CONTROLS KEYWORDS
#Defaults are used for keywords not listed here
  tolpar=0.01
  tolsosc=0.1
  maxiter=10
  maxchange=1.5
  MaxchangeRealm=regression
END REG GN CONTROLS
```

Model_Command_Lines Input Block: Control Execution of the Process model (required)

The Model_Command_Lines input block defines the command needed to execute a process model. Different commands can be defined to execute the process model in a way that produces one forward run, a run that also produces sensitivities (here the use of 'der' is derived from sensitivities being equivalent derivatives). For completeness, the option in which the process model produces only sensitivities is also included, but this option is rarely used.

The three keywords are listed below. Keywords are not case sensitive.

Command — Operating system command that executes the process model(s). Up to 2,000 characters. The command can be the name of an executable file or an absolute or relative pathname. If the command includes spaces, it needs to be enclosed in single quotes; otherwise, quotes are optional. If the command line includes single quotes, a way to run it without single quotes needs to be implemented. For example, on a Windows operating system it could be placed in a batch file. There is no default.

Purpose — The type of process-model run executed by Command. For the three options described here, the term "simulated values" can refer to simulated equivalents to observations, predictions, or both. Default=forward.

forward: The command makes a model run that generates simulated values.

derivatives: The command makes a model run that generates sensitivities of simulated values with respect to specified parameters.

forward&der: The command makes a model run that generates both simulated values and sensitivities of the simulated values with respect to specified parameters.

CommandID — A name for the command. The command name is used at the top of the main output file in a list of the programs run; it does not influence the execution process. There is no default.

If Blockformat KEYWORDS is selected by designation or default, the Command keyword needs to be listed first, followed by the two other keywords in any order. The Command keyword needs to be the first keyword on a new line. This sequence is repeated for each command. Each purpose can only be listed once.

Blockformat TABLE requires COLUMNLABELS because there is no default order for these variables.

Example of a Model_Command_Lines input block:

```
BEGIN MODEL COMMAND LINES
# Need single quotes around 'Command=value' if it includes
# any spaces. Quotes are optional otherwise
  Command='m.bat'
  purpose=forward&der
  CommandId=modflow
END MODEL COMMAND LINES
```

In many situations the 'process model' is actually a sequence of models. This is accommodated using UCODE_2005 by assembling the required simulations in a batch file on a Windows operating system, script on a Linux operating system, and so on. Then the batch file or script is defined using the keyword Command.

For example, on a Windows operating system the batch file listed in the Command_Lines input block might contain the following.

```
call ..\..\test-data-win\data-obs\tc1-obs.bat
call ..\..\test-data-win\data-preds\tc1-pred.bat
```

The listed batch files need to be inspected carefully. For example, any line containing the command 'pause' in a batch file causes execution to halt until a key on the keyboard is pressed. This would require constant attention as the UCODE_2005 run proceeded.

Chapter 7: INPUT TO DEFINE PARAMETERS

Up to four input blocks can be used to define parameters: Parameter_Groups, Parameter_Data, Parameter_Values, and Derived_Parameters. The second is always needed; the other three are optional.

Quantities needed to define parameters include the starting parameter value, values that govern how perturbation sensitivities are calculated, and so on. Some of these quantities may be the same for many parameters, and it is convenient to define such quantities for the parameters as a group. UCODE_2005 provides for this using the optional Parameter_Groups input block.

Information specific to individual parameters is defined in the required Parameter_Data input block. If differences occur, data specified in the Parameter_Data input block replace data specified in the Parameter_Groups input block.

To start with a set of parameter values previously calculated by UCODE_2005, or any alternative values, while preserving a record of the original starting values, use the Parameter_Values input block.

Commonly some parameters represent quantities that need to be manipulated before being written into the process-model input file. For example, it may be useful to define the parameter as hydraulic conductivity of the riverbed while the model input requires a conductance defined as the hydraulic conductivity times area divided by riverbed thickness. Or, it may be useful to define a recharge parameter in millimeters per year within a model for which the length unit is meters. To calculate model inputs that are a function of defined parameters, use the Derived_Parameters input block.

Parameter_Groups Input Block (optional)

Use the Parameter_Groups input block to assign data that apply to all or many of the parameters within defined groups. Data for individual parameters can be assigned in the subsequently read Parameter_Data input block, and, when quantities specified in the Parameter_Groups block are repeated in the Parameter_Data block, the data specified in the Parameter_Data block are used.

Keywords in this input block include:

GroupName   - The name of the group (up to 12 characters; not case sensitive). Default=ParamDefault

Other keywords   - Any keyword from the Parameter_Data input block.

If Blockformat KEYWORDS is selected by designation or default, keywords associated with a parameter group in the Parameter_Groups input block need to be grouped together and follow the related GroupName. The GroupName keyword needs to be the first keyword on a new line.

If Blockformat TABLE format is selected, COLUMNLABELS are needed because there is no default column order for the Parameter_Groups input block.

Example of a Parameter_Groups input block:

```
BEGIN PARAMETER GROUPS KEYWORDS
   groupname=Default   adjustable=yes tolpar=0.005
END PARAMETER GROUPS
```

Parameter_Data Input Block (required)

The Parameter_Data input block provides information about individual parameters.

For each parameter, keywords listed below can be defined either in the Parameter_Data or the Parameter_Group input block except that ParamName needs to be specified here.

By specifying the groupname for a parameter in this block, all values associated with that group in the Parameter_Groups block are assigned to the parameter. Any data defined for a parameter in this input block overrides data from the Parameter_Groups input block.

Linking parameters such that multiple parameters defined in the Parameter_Data input block are treated as if they are one parameter can be useful in many situations. For example, if parameters are too insensitive to be estimated individually, linking them together can produce a parameter that can be estimated. Also, linking parameters is often needed to obtain useful parameter correlation coefficients (Hill and Østerby, 2003). In UCODE_2005, parameters can be linked by making suitable changes in the Parameter_Data input block and using the Derived_Parameters input block. The methodology is described in the documentation of the Derived_Parameters input block.

The KEYWORDS for the Parameter_Data input block are:

ParamName - Parameter name (up to 12 characters; not case sensitive) – a character string that is used in a template file or in an equation of a derived parameter in the Derived_Parameters input block. Each parameter name needs to be unique and can not be the same as any parameter name defined in the Derived_Parameter input block.

Naming convention for ParamName:

1) The first character needs to be a letter of the set (A-Z, a-z); and

2) All remaining characters need to be a letter, digit, or member of the set:

_ . : & # @ (underscore, dot, colon, ampersand, number sign, at symbol).

The restrictions are needed for the parameter names to be used in the equations defined in Chapter 13.

GroupName - Group name (up to 12 characters; not case sensitive). Each parameter needs to be a member of one group.
Default=ParamDefault

StartValue - Starting parameter value. Default=A huge real number. The huge real number is obtained for the computer being used and commonly is about 10^{38}.

The following two keywords define lower and upper reasonable values for the parameter; they do not constrain the parameter value. Estimates that are not reasonable can be important indicators of model error (Poeter and Hill, 1996; Hill and Tiedeman, in press, Chapter 5.5). Estimated parameter values that are outside the defined range are identified in the main output file and in the _pc data-exchange file. Estimated parameter values for which the entire linear confidence interval is outside the defined range also are identified, and are a strong indicator of model error.

LowerValue — Smallest reasonable value for this parameter. Default= −(Huge real number). In absolute value, commonly about 10^{38}.

UpperValue — Largest reasonable value for this parameter. Default= +(Huge real number). Commonly about $+10^{38}$.

The following keyword is used only when Optimize=yes in the UCODE_Control_Data input block. In that situation, Constrain=yes indicates that the estimated parameter value is required to remain between LowerConstraint and UpperConstraint. Using constraints to avoid unreasonable parameter values can diminish a valuable tool for identifying model error. Using constraints to avoid parameter values that prevent a code from running (for example, n>1 for a vanGenuchten characteristic function) can be important to achieving successful regressions.

Constrain — yes: constrain parameter values using LowerConstraint and UpperConstraint. Default=no.

The following two keywords serve one of two purposes depending on keywords Optimize and SOSSurface in the UCODE_Control_Data input block.
  (1) When Optimize=yes, Constrain=yes indicates that the estimated parameter value is required to remain between LowerConstraint and UpperConstraint.
  (2) When SOSSurface=yes, LowerConstraint and UpperConstraint are used with SOSIncrement to define sets of parameter values.

LowerConstraint — Lower limit of considered parameter values.

UpperConstraint — Upper limit of considered parameter values.

When Optimize=yes and Constrain=yes, constraining parameter values is implemented as follows. If a parameter value has been constrained and is about to violate a specified constraint in a given parameter-estimation iteration, the step size for the parameter-estimation iteration is limited to the amount needed for the parameter to reach the limiting constraint.

If one or more parameters have been omitted from the regression because of constraints, in each subsequent parameter-estimation iteration their sensitivities are re-evaluated and regression is attempted with the parameters. Parameter values calculated to move to within the constraints are allowed to change with the other parameter values. Parameter values calculated to move outside the constraints are considered as follows. First the parameter with the largest fractional change is omitted and the parameter change vector is recomputed. If additional parameter values move outside the constraints, the procedure is repeated until a set of parameter values is obtained that are calculated to be within their constraints.

Adjustable - yes: change this value as needed depending on the purpose of the UCODE_2005 run defined in the UCODE_Control_Data input file. no: leave the value of this parameter unchanged. Default=no.

PerturbAmt - Fractional amount of parameter value to perturb to calculate sensitivity. Commonly 0.01 to 0.10. Default=0.01. See discussion in Chapter 3.

Transform - yes: log-transform the parameter for the regression. no: estimate the native value in the regression. If Transform=yes, any transformed values printed to files are in log base 10 except that weighted residuals for prior information are in natural log. Within the program calculations are done using natural logs. Default=no.

TolPar - Replaces, for this parameter, the value of TolPar from the Reg_GN_Controls input block. Default=TolPar from the Reg_GN_Controls input block.

MaxChange - Maximum fractional parameter change allowed between parameter iterations. Default=2.0.

SenMethod - A flag indicating how sensitivities are obtained. Sensitivities for different parameters can be obtained using different methods. For each parameter, sensitivities for all simulated values are calculated by a single method. Options include:

SenMethod	Source of sensitivities
-1	Process-model output file. Sensitivities for log transformed parameters are read as transformed sensitivities.
0	Process-model output file. Sensitivities for log transformed parameters are read as native parameter sensitivities and transformed by UCODE_2005.
1	Calculate by forward-difference perturbation. (Default)
2	Calculate by central-difference perturbation (two-point method)

ScalePval - A positive number used in scaled sensitivities if the absolute value of the parameter value is less than ScalePval. If the absolute value of the parameter value is less than ScalePval, ScalePval replaces the parameter value in the scaling of one-percent, dimensionless, and composite scaled sensitivities. Default=StartValue/100.

Good choices for ScalePval are (1) the smallest (in absolute value) reasonable value of the parameter or (2) a value two to three orders of magnitude smaller than the expected value, which is typically the starting value. If the smallest reasonable value is 0.0, a reasonable non-zero value needs to be used. ScalePval has no effect on the scaled sensitivities for log-transformed parameters or on parameter estimation.

SOSIncrement   - The number of values to be considered between and including the specified LowerConstraint and UpperConstraint for parameters with adjustable=yes. Only used when SOSsurface=yes in the UCODE_Control_Data input block. The number of runs equals the product of all non-zero values of SOSIncrement for parameters with adjustable=yes. A value of 5 results in parameter values at the LowerConstraint, the UpperConstraint and three evenly spaced intermediate values. For log-transfomed parameters, the intermediate values are evenly spaced in log space. Default=5.

NonLinearInterval – yes: calculate nonlinear intervals for this parameter when NonlinearIntervals=yes in the UCODE_CONTROL_DATA block. no: do not calculate nonlinear intervals for this parameter. Default=no.

If Blockformat KEYWORDS is selected by designation or default, keywords defining a parameter in the Parameter_Data input block need to be grouped together and follow the related ParamName. The ParamName keyword needs to be the first keyword on a new line. ParamName and associated keywords are repeated to define multiple parameters.

If blockformat TABLE is selected without indicating ColumnLabels, the default column order is the order in which the keywords are defined above and listed below. No columns are ignored and a column for each keyword is needed. If ColumnLabels are indicated, the column labels can appear in any order; the ParamName keyword need not be first, though it often is first. If keyword NonLinearInterval is used, column labels are needed because it is not included in the default column order.

Default Column Order: PARAMNAME GROUPNAME STARTVALUE LOWERVALUE UPPERVALUE CONSTRAIN LOWERCONSTRAINT UPPERCONSTRAINT ADJUSTABLE PERTURBAMT TRANSFORM TOLPAR MAXCHANGE SENMETHOD SCALEPVAL SOSINCREMENT NONLINEARINTERVAL

Example of a Parameter Data input block:

```
BEGIN PARAMETER DATA TABLE
# Defaults are used for keywords that are not listed.
# Selected values are listed here for easy reference:
# SenMethod=1 (sensitivities are calculated by perturbation),
# maxchange=2.0
nrow=9  ncol=6 columnlabels  groupname=mypars
paramname startvalue lowervalue uppervalue scalepval   adjustable
Wells TR   -1.1000      -1.4        -0.8      1.0E-3     yes
RCH Zone 1   60.        30.0        80.0      1.0E-2     yes
RCH Zone 2   30.        20.0        60.0      1.0E-2     yes
Rivers     .00120      .00012      1.2E-2     1.0E-6      no
SS 1       .00130      .00013      1.3E-2     1.0E-6     yes
HK 1       .00030      .00003      3.0E-3     1.0E-7     yes
Vert_K CB  1.0E-7      1.0E-8      1.0E-6     1.0E-10    yes
SS_2        6.2E-5      2.0E-5      2.0E-3     1.0E-7      no
HK 2        4.0E-5      4.0E-6      4.0E-4     1.0E-8      no
END PARAMETER DATA
```

Parameter_Values Input Block: Use Alternative Starting Parameter Values (optional)

A Parameter_Values input block can optionally follow the Parameter_Data block. A Parameter_Values block provides a convenient way to use the following.

(1) Parameter values generated by an earlier run of UCODE_2005 and written to data-exchange file fn._pasub (fn is a filename prefix defined on the command line).

(2) Parameter values generated by an external preprocessing program.

(3) Alternate starting values without losing a record of starting values listed in the Parameter_Groups or Parameter_Data input blocks. This can be important because regression performance sometimes can be improved by using alternate starting values (Hill and Tiedeman, in press).

All parameters need to have a starting value defined at least once using the StartValue keyword in the Parameter_Groups, Parameter_Data, or Parameter_Values input blocks. If defined more than once, values defined in the Parameter_Values input block replace values defined elsewhere; values in the Parameter_Data input block replace values defined in the Parameter_Groups input block.

The KEYWORDS for the Parameter_Values input block are:

ParamName   - The name of the parameter for which a value is specified.

StartValue   - The specified parameter value.

If Blockformat KEYWORDS is selected by designation or default, each keyword ParamName needs to be followed by the associated keyword StartValue. The ParamName keyword needs to be the first keyword on a new line.

If Blockformat TABLE is selected without indicating COLUMNLABELS, the default column order is as listed above: `ParamName StartValue.`

Example of the Parameter Values input block:

```
BEGIN Parameter Values TABLE
# These values override values listed in Parameter Data
# input block
  nrow=9  ncol=2  columnlabels
  paramname  startvalue
  Wells TR    -1.1000
  RCH Zone 1  6.3072E+1
  RCH Zone 2  3.1536E+1
  Rivers      1.2000E-3
  SS 1        1.3000E-3
  HK 1        3.0000E-4
  Vert K CB   1.0000E-7
  SS 2        2.0000E-4
  HK_2        4.0000E-5
END Parameter_Values
```

Derived_Parameters Input Block: Define Model Inputs as Functions of Parameters (optional)

A Derived_Parameter input block can optionally end the parameter input section. The keywords of the Derived_Parameters input block are:

DerParName — Name of derived parameter (up to 12 characters, not case sensitive) – a character string that is to be substituted into a process-model input file or used in another equation. Each derived parameter name needs to be unique and can not be the same as any parameter name defined in the Parameter_Data input block

Naming convention for ParamName:

1) The first character needs to be a letter of the set (A-Z, a-z); and

2) All remaining characters need to be a letter, digit, or member of the set:

_ . : & # @ (underscore, dot, colon, ampersand, number sign, at symbol).

The restrictions are needed for the parameter names to be used in the equations defined in Chapter 13.

DerParEqn — An equation without an "equal" sign (that is, just the equation right-hand side) by which the derived parameter is calculated, generally using defined parameters.

The following rules apply to DerParEqn:
1) Equation protocols are outlined in the EQUATION section of this document in Chapter 13.
2) Variables listed in an equation can be a parameter name from the Parameter_Data section, or a derived parameter name that is previously defined in the Derived_Parameters input block.
3) Template files can be constructed using derived parameter names defined in the Derived_Parameters input block or parameter names defined in the Parameter_Data input block. Not all defined names need to appear in the template files.

If Blockformat KEYWORDS is selected by designation or default, each keyword DerParName needs to be followed by the associated DerParEqn. The DerParName keyword needs to be the first keyword on a new line.

If Blockformat TABLE is selected without indicating COLUMNLABELS, the default column order is as listed above: `DerParNamE DerParEqn`.

Example of the Derived Parameters input block:

```
BEGIN Derived Parameters TABLE
nrow=1 ncol=2 columnlabels
derparname derpareqn
K          T/b
END Derived Parameters
```

As mentioned in the documentation of the Parameter_Data input block, linking parameters can be useful in many situations. In UCODE_2005, parameters can be linked as follows.

1. Identify a parameter to which other parameters are linked. This is called the anchor parameter.
2. Delete the linked parameters from the Parameter_Data input block. The template files remain unchanged.
3. In the Derived_Parameters input block, define the linked parameters using the anchor parameter. Use equations as needed to maintain the desired relation between the linked parameters and the anchor parameter.

For example, consider the following Parameter Data input block

```
BEGIN PARAMETER_DATA TABLE
# Use the defaults:
# SenMethod=1 (sensitivities are calculated by perturbation),
# maxchange=2.0
# tolpar from the Reg GN Controls input block is used for all
# parameters, perturbamt=0.01
#    1         2         3         4         5         6
nrow=5 ncol=6 columnlabels  GroupName=MyPars
paramname STARTVALUE  lowervalue uppervalue     scalepval transform
RCH Zone 2 3.1536E+1    20.0       60.0         1.0E-2    no
Rivers     1.2000E-3    1.2E-4     1.2E-2       1.0E-6    yes
HK 1       3.0000E-4    3.0E-5     3.0E-3       1.0E-7    yes
Vert K CB  1.0000E-7    1.0E-8     1.0E-6       1.0E-10   yes
HK 2       4.0000E-5    4.0E-6     4.0E-4       1.0E-8    yes
END PARAMETER DATA
```

Parameters HK_1, HK_2, and Rivers can be combined with the following Parameter_Data and Derived_Parameters input blocks, using HK_1 as the anchor parameter.

```
BEGIN PARAMETER DATA TABLE
# Use the defaults: SenMethod=1 (sensitivities are calculated by
perturbation), maxchange=2.0, tolpar from the Reg GN Controls input
block is used for all parameters, perturbamt=0.01
# Combined parameters are saved here in comment statements:
#Rivers     1.2000E-3    1.2E-4     1.2E-2       1.0E-6    yes
#HK_2       4.0000E-5    4.0E-6     4.0E-4       1.0E-8    yes
#    1         2         3         4         5         6
nrow=3 ncol=6 columnlabels  GroupName=MyPars
paramname STARTVALUE  lowervalue uppervalue     scalepval transform
RCH Zone 2 3.1536E+1    20.0       60.0         1.0E-2    no
HK 1       3.0000E-4    3.0E-5     3.0E-3       1.0E-7    yes
Vert K CB  1.0000E-7    1.0E-8     1.0E-6       1.0E-10   yes
END PARAMETER DATA
```

```
BEGIN Derived Parameters TABLE
nrow=2 ncol=2 columnlabels
derparname derpareqn
Rivers      HK_1*1.2000E-3/3.0000E-4
HK_2        HK_1*4.0000E-5/3.0000E-4
END Derived Parameters
```

In this case, the equations in the Derived_Parameter input block ensure that the linked parameters maintain a constant ratio to the value of HK_1 as it is adjusted. That is, the ratio of each linked parameter value to the anchor parameter value stays the same as it was for the starting parameter values. If another relation is desired, such as preserving a sum or difference, the equations can be changed accordingly.

Chapter 8: INPUT TO DEFINE OBSERVATIONS AND PREDICTIONS

Observations and predictions are defined using input blocks that are nearly identical, so both are described in this chapter.

UCODE_2005 allows substantial flexibility in the values used as simulated equivalents of the observations or as predictions. Such values can be calculated from one or more of the values read from the process-model output file(s). This flexibility is needed, for example, when an observation is located at a point in space that is not represented by a printed value in the simulation output, but rather falls between the locations of a number of printed values. In this situation, UCODE_2005 can be instructed to interpolate the printed values to obtain a simulated value for comparison with the observation. Alternatively, an observation or prediction may be the equivalent of the sum of many values or portions of values printed by the application code(s). In UCODE_2005, these values or appropriate portions of them can be summed to obtain an equivalent simulated value or prediction using the Derived_Observations and Derived_Predictions input blocks.

Observations

Up to three input blocks can be used to define observations: Observation_Groups, Observation_Data, and Derived_Observations. To define observations, the second input block is always needed; the first and last are optional.

Observations need to be defined for all UCODE_2005 modes except the prediction mode, for which all observation input blocks need to be omitted. The prediction mode is defined when prediction=yes in the UCODE_Control_Data input block. For mode definition, see table 3 and Chapter 17.

The quantities needed to define observations include the observed value, values that quantify the accuracy of the observed value, and so on. Some of these quantities are the same for many observations, and it is convenient to define them for observations as a group. UCODE_2005 provides for this using the optional Observation_Groups input block.

Information specific to individual observations is defined in the required Observation_Data input block. If the information differs from that specified in the Observation_Groups input block, the data specified in the Observation_Data input block are used.

In some circumstances the observation represents a quantity that is not directly provided in the process-model output file. For example, hydraulic-heads may be produced by the process model, but the observation may be drawdown. In these circumstances it is advantageous to calculate simulated equivalents that are a function of values read from the output file(s). This can be accomplished by using Derived_Observations, which can be specified in the Observation_Data input block or, if the user prefers to separate these for organizational reasons, they can be specified in the Derived_Observations input block.

Observation_Groups Input Block (optional)

Use the Observation_Groups input block to define groups and to assign data that apply to all or many of the observations within defined groups. Data for individual observations can be assigned in the subsequently read Observation_Data input block. When quantities are specified in both blocks, data specified in the Observation_Data block are used. Keywords in this input block include:

GroupName — Name for a group of observations (up to 12 characters; not case sensitive). Default=DefaultObs.

UseFlag — yes: use the simulated values in this group to compare against observed values in the regression. no: do not include the observations of this group in the regression. Use 'no' for items only read to calculate derived observations. Default=yes.

PlotSymbol — An integer intended for use in post-processing programs to assign symbols for plotting. Default=1.

WtMultiplier — Value used to multiply the weights associated with members of a group when the weighting is defined using Statistic and StatFlag keywords described for the Observation_Data input block. Default=1.0.

A CovMatrix is needed to represent correlation between errors in the members of a group. One matrix is used to define the weighting for all the members in the group. Members with independent errors have zero off-diagonal terms in the matrix.

CovMatrix — Name of the error variance-covariance matrix. The matrix is specified in the Matrix_Files input block of Chapter 10. CovMatrix should not be specified if there is no correlation between errors in the members of a group.

Other keywords — Any keyword from the Observation_Data input block.

If Blockformat KEYWORDS is selected by designation or default, keywords associated with an observation group need to be grouped together and follow the related GroupName. The GroupName keyword needs to be the first keyword on a new line.

If Blockformat TABLE format is selected, COLUMNLABELS are needed because there is no default column order for the Observation_Groups input block.

Example of an Observation Groups input block:

```
BEGIN OBSERVATION GROUPS TABLE
  nrow=5 ncol=3  columnlabels statflag=sd
  GROUPNAME   plotsymbol  useflag
  heads        1           yes
  headsext     0            no
  hds4der      1           yes
  notused      0            no
  flowobs      2           yes
END OBSERVATION GROUPS
```

Observation_Data Input Block (required except for prediction mode)

The Observation_Data input block provides information about individual observations. For each observation, all the keywords listed below need to be defined either in the Observation_Data or the Observation_Groups input block or by default.

By specifying a groupname for an observation, all definitions for that group from the Observation_Groups input block are assigned to the observation. Keywords defined in the Observation_Groups input block only need to appear here to change the designation. Any data defined for an observation in this block overrides data from the Observation_Groups input block with one exception: if a CovMatrix is specified for the group in the Observation_Groups input block, then values assigned to Statistic and StatFlag in the Observation_Data input block are ignored.

The following keywords are available for this input block:

ObsName   - Observation name (up to 20 characters, not case sensitive). Each observation name needs to be unique.

Naming convention for ObsName:

1) The first character needs to be a letter of the set (A-Z, a-z); and

2) All remaining characters need to be a letter, digit, or member of the set:

_ . : & # @ (underscore, dot, colon, ampersand, number sign, at symbol).

3) The name 'dum' can not be used.

The restrictions are needed for the parameter names to be used in the equations defined in Chapter 13.

ObsValue   - Observation value or, if the observation is below a detection limit (NonDetect=yes; see fifth keyword down), the value to use in the regression.

Statistic   - Value used to calculate the observation weight.

StatFlag   - Character string that defines the corresponding statistic and how it is used to calculate the weight. No default. Options are:

StatFlag	Statistic	Weight calculated as
VAR	Variance	$1/\text{Statistic}$
SD	Standard deviation	$1/(\text{Statistic})^2$
CV	Coefficient of variation	$1/(\text{Statistic} \times \text{ObsValue})^2$
WT	Weight	Statistic
SQRWT	Square root of the weight	Statistic^2

GroupName   - Group name from the Observation_Groups input block. The group attributes defined in the Observation_Groups input block are assigned to the observation and are then changed to attributes from the Observation_Data input block if specified. If the groupname used here has not been defined, the observation will not be used in the regression. Default=DefaultObs.

Equation   - An equation without an "equal" sign that defines how to calculate a simulated equivalent value to compare to the observation from simulated equivalents of previously defined observations. Use of _ for the equation indicates that the extracted value is used directly and it is not considered to be derived. Default= _.

See the comment under the Derived_Observations section concerning calculating sensitivities for derived observations. The following rules apply to equations:

1. Equation protocols are outlined in Chapter 13 of this report. Equations can contain spaces only if the equation is enclosed in double or single quotes.

2. Equations can cite any ObsName previously listed in the Observation_Data input block. Cited observations can have a UseFlag of "yes" or "no".

3. Equations can be nonlinear.

4. An observation derived with an equation can not be used in conjunction with sensitivities calculated by the process model (SenMethod=-1 or 0 in the Parameter_Data input block). Rather sensitivities need to be calculated by perturbation (SenMethod=1 or 2).

Nondetects are common when using concentration and other types of observations, and can be important to include in any analysis, as discussed by Helsel (2004). That text discusses the disadvantages of the commonly used methods of deletion and substitution. Deleting non-detects can bias the solutions to higher values; substitution using detection limits (commonly one-half the detection limit is used) can bias the solution because of the generally downward trend of detection limits. These common methods can be pursued using UCODE_2005, but the results need to be considered carefully.

NonDetect   - yes: the observed value is below a detection limit and the value specified as ObsValue is used in the regression. This is what Helsel (2004) calls a substitution method. If NonDetect=yes, calculation of the residual depends on the simulated value: if simulated value $\leq$ ObsValue, the residual equals zero; if simulated value > ObsValue, the residual equals ObsValue minus the simulated value, as usual. If NonDetect=yes, default columns can not be used and the runs statistic is not printed. Default=no.

The next keyword controls the use of simulated values to calculate weights as described in equation 3 of Chapter 3. The calculation uses keyword TolParWtOS from the Reg_GN_Control input block. If WtOSConstant>0, the following apply: (1) default column order cannot be used and columnlabels are needed and (2) StatFlag=CV is needed for this observation and a correlated weight matrix cannot be used for the group. Generally the value of WtOSConstant is the same for all members of a group.

WtOSConstant   – The constant η in equation 3. Default=0.0.

If Blockformat KEYWORDS is selected by designation or default, keywords defining an observation in the Observation_Data input block need to be grouped together and follow the related ObsName. The ObsName keyword needs to be the first keyword on a new line. ObsName and associated keywords are repeated to define multiple observations.

If blockformat TABLE is selected without indicating ColumnLabels, the default column order listed below is used. No columns are ignored and a column for each keyword is expected. If ColumnLabels are indicated, the column labels can appear in any order; the ObsName keyword need not be first, though it often is. The keywords WtOSConstant and NonDetect are not included in the default column order. If WtOSConstant and(or) NonDetect need to be other than the 0.0 default value, they need to be defined in the Observation_Groups input block or column labels need to be used.

Default Column Order:
```
OBSNAME OBSVALUE STATISTIC STATFLAG GROUPNAME EQUATION
```

Example of the Observation_Data input block. This example is consistent with the example in the Observation_Groups input block example in the previous section.
```
BEGIN OBSERVATION_DATA FILES
tc1.hed
tc1.flo
END OBSERVATION DATA
```

Files tc1.hed and tc1.flo are read. For example, file tc1.flo:
```
#For GroupName=flowobs, StatFlag=sd in
#the Observation Groups input block
BEGIN OBSERVATION DATA TABLE
  NROW=5   NCOL=5   COLUMNLABELS
  Obsname    obsvalue statistic  equation                    groupname
  flow.ss     -4.4     0.4                                    flowobs
  flow.t3     -4.1     0.38                                   notused
  flow.t12    -2.2     0.21                                   notused
  flow.t3 ss   0.3     0.55      "flow.t3 - flow.ss"          flowobs
  flow.t12 ss  2.2     0.45      "flow.t12 - flow.ss"         flowobs
END OBSERVATION DATA
```

In file tc1.hed, which is listed next, the equations demonstrate some available capabilities. For example, h1.0 and h1.0a may be heads in layers intersected by a single well so that the simulated equivalent needs to reflect both values. Here, a simple average is chosen and given the name h1.0_der. The h2.0 series is presented to display some of the equation functionality; h2.0_der just equals h2.0a.

```
BEGIN OBSERVATION DATA TABLE
#StatFlag designation here overrides value from the
#Observation Groups input block
NROW=8  NCOL=6  COLUMNLABELS
obsname   obsvalue    statistic statflag  equation          GROUPNAME
h1.0          0.0       1.0025   var          _             headsext
h1.0a         0.0       1.0025   var          _             headsext
h1.0 der  101.8040      1.0025   var   (h1.0+h1.0a)/2.      hds4der
h1.1        -0.290E-01  0.0025   var                        heads
h1.12       -0.129      0.0025   var                        heads
h2.0      128.1170      1.0025   var                        headsext
h2.0a     138.2260      1.0025   var                        headsext
h2.0 der  128.1170      1.0025   var sqrt(h2.0^2)+h2.0a-h2.0 hds4der
END OBSERVATION_DATA TABLE
```

Derived_Observations Input Block: Define Simulated Equivalents as Functions of Model Outputs (optional)

The Derived_Observations input block is identical to the Observation_Data input block except in name. It is included in UCODE_2005 so the user can define derived observations in a separate block, which is convenient in some circumstances.

Derived_Observations can not be used in conjunction with sensitivities calculated by the process model (SenMethod=-1 or 0 in the Parameter_Data input block). Rather, if the quantity of interest cannot be calculated by the process model, sensitivities for derived observations need to be calculated by perturbation (SenMethod=1 or 2).

Example of the Derived_Observations input block. ObsNames head1 and head6 need to be defined in the Observation_Data block because they are not defined previously in the Derived_Observations block:

```
BEGIN Derived Observations TABLE
nrow=1 ncol=6 columnlabels
obsname   obsvalue   statistic   statflag   equation     GROUPNAME
dd5        0.012      1e-6        VAR        head1-head6  drawdowns
END Derived  Observations
```

Predictions

Up to three input blocks can be used to define predictions: Prediction_Groups, Prediction_Data, and Derived_ Predictions. To define predictions, the second input block is always needed; the first and last are optional.

Predictions need to be defined for three UCODE_2005 modes: prediction mode (table 3) and advanced-test-model-linearity and, usually, nonlinear-uncertainty mode (Chapter 17). For other modes, all prediction input blocks need to be omitted.

The modes are defined with keywords in the UCODE_Control_Data input block. For mode definition, see table 3 and Chapter 17.

The prediction input blocks can be omitted by deleting them or placing a "#" character in the first place of each line of the blocks.

Up to three input blocks can be used to define predictions: Prediction_Groups, Prediction_Data, and Derived_Predictions. The second is always needed; the first and last are optional.

The quantities needed to define predictions include a reference value, a scaling factor that can be used to calculate scaled prediction sensitivities, and so on. Some of these quantities are the same for many predictions, and it is convenient to define them for predictions as a group. UCODE_2005 provides for this using the optional Prediction_Groups input block.

Information specific to individual predictions is defined in the required Prediction_Data input block. If differences occur, data specified in the Prediction_Data input block replace data specified in the Prediction_Groups input block.

In some circumstances the prediction represents a quantity that is not directly provided in the process-model output file. For example, hydraulic-heads may be produced by the process model, but the prediction may be drawdown. In these circumstances it is advantageous to calculate a prediction that is a function of values read from process-model output file(s). This can be accomplished by using derived predictions, which can be specified using equations in the Prediction_Data input block or, if the user prefers to separate these for organizational reasons, in the Derived_ Predictions input block. Derived predictions can not be used in conjunction with sensitivities calculated by the process model (SenMethod=-1 or 0 in the Parameter_Data input block).

Prediction_Groups Input Block (optional)

Use the Prediction_Groups input block to define groups and to assign data to all predictions within a group. Data for individual predictions can be assigned in the subsequently read Prediction_Data input block. When quantities are specified in both input blocks, the data specified in the Prediction_Data input block are used. Keywords in this input block include:

GroupName — Name for a group of predictions (up to 12 characters; not case sensitive). GroupNames for predictions need o be unique and different than any listed in the Observation_Groups input block. Default=DefaultPreds.

UseFlag — yes: report and analyze the predictions in this group. no: do not report and analyze the predictions in this group. Use 'no' for items extracted only to calculate derived predictions. Default=yes.

PlotSymbol — An integer intended for use in post-processing programs to assign symbols for plotting. Default=1.

Other keywords — Any keyword from the Prediction_Data input block.

If Blockformat KEYWORDS is selected by designation or default, keywords associated with a prediction group need to be grouped together and follow the related GroupName. The GroupName keyword needs to be the first keyword on a new line.

If Blockformat TABLE format is selected, COLUMNLABELS are needed because there is no default column order for the Prediction_Groups input block.

Example of a Prediction_Groups input block:

```
BEGIN PREDICTION GROUPS KEYWORDS
groupname= heads
  plotsymbol=1 useflag=yes
groupname=flows
  plotsymbol=2 useflag=yes
END PREDICTION GROUPS
```

Prediction_Data Input Block (required only for prediction mode)

The Prediction_Data input block provides information about individual predictions. For each prediction, the keywords listed below need to be defined either by default or in the Prediction_Data or the Prediction_Groups input block.

By specifying a groupname for a prediction, all definitions for that group from the Prediction_Groups input block are assigned to the prediction. Keywords defined in the Prediction_Groups input block only need to appear here to change the designation. Any data defined for a prediction in this block overrides data from the Prediction_Groups input block.

The following keywords are available for this input block:

PredName — Prediction name (up to 20 characters, not case sensitive). Each prediction name needs to be unique.

Naming convention for PredName:

1) The first character needs to be a letter of the set (A-Z, a-z); and

2) All remaining characters need to be a letter, digit, or member of the set:

_ . : & # @ (underscore, dot, colon, ampersand, number sign, at symbol).

3) The name 'dum' can not be used.

The restrictions are needed for the parameter names to be used in the equations defined in Chapter 13.

RefValue — Reference value to which the prediction is compared. This value is used to calculate scaled sensitivities. If RefValue equals zero, the scaled sensitivities are set to zero.

MeasStatistic — A statistic used to calculate the variance with which the predicted quantity could be measured.

The variance is printed in the _pv data-exchange file, which is used by LINEAR_UNCERTAINTY (Chapter 15) and the nonlinear-uncertainty mode of UCODE_2005 (Chapter 17) to calculate prediction intervals (Chapter 3).

MeasStatFlag — Character string that defines how the corresponding MeasStatistic is used to calculate the measurement error for the prediction. No default. Options are:

MeasStatFlag	Variance is calculated as
VAR	MeasStatistic
SD	$(MeasStatistic)^2$

GroupName — Group name from the Prediction_Groups input block. The group attributes defined in the Prediction_Groups input block are assigned to the prediction and are then changed to attributes specified in the Prediction_Data input block if applicable.

Equation   - An equation without an "equal" sign that defines how to calculate a derived prediction from previously defined PredNames. Use of _ for the equation indicates that the extracted value is used directly, so it is not considered to be derived. Default= '_'.

See the comment under the Derived_Predictions section concerning calculating sensitivities for derived predictions. The following rules apply to equations:

1. Equation protocols are outlined in Chapter 13 of this report. Equations can contain spaces only if the equation is enclosed in double or single quotes.

2. Equations can cite any PredName previously listed in the Prediction_Data input block. Cited predictions can have a UseFlag of "yes" or "no".

3. Equations can be nonlinear.

4. A prediction derived with an equation can not be used in conjunction with sensitivities calculated by the process model (SenMethod=-1 or 0 in the Parameter_Data input block). Rather, sensitivities need to be calculated by perturbation (SenMethod=1 or 2).

If Blockformat KEYWORDS is selected by designation or default, keywords defining a prediction in the Prediction_Data input block need to be grouped together and follow the related PredName. The PredName keyword needs to be the first keyword on a new line. PredName and associated keywords are repeated to define multiple predictions.

If blockformat TABLE is selected without indicating ColumnLabels, the default column order is used and is listed below. No columns are ignored and a column for each keyword is expected. If ColumnLabels are indicated, the column labels can appear in any order; the PredName keyword need not be first, though it often is.

Default Column Order:
```
PREDNAME REFVALUE MEASSTATISTIC MEASSTATFLAG GROUPNAME EQUATION
```

Example Prediction_Data input block:
```
BEGIN PREDICTION_DATA FILES
tc1.hed
tc1.flo
END PREDICTION_DATA
```

Files tc1.hed and tc1.flo are read. For example, file tc1.flo:

```
#For GroupName=FlowPred, MeasStatFlag=SD in
#the Prediction Groups input block
#Double quotes are needed for the equations because they
#include spaces.
BEGIN PREDICTION DATA TABLE
  NROW=5  NCOL=5  COLUMNLABELS
  predname refvalue measstatistic  equation          groupname
  flow.ss    -4.4      0.4                            flowpred
  flow.t3    -4.1      0.38                           notused
  flow.t12   -2.2      0.21                           notused
  flow.t3 ss  0.3      0.55      "flow.t3 - flow.ss"  flowpred
  flow.t12 ss 2.2      0.45      "flow.t12 - flow.ss" flowpred
END PREDICTION DATA
```

Derived_Predictions Input Block: Define Predictions as Functions of Model Outputs (optional)

The Derived_Predictions input block is identical to the Prediction_Data input block except in name. It is included in UCODE_2005 so the user can define derived predictions in a separate block, which may be convenient in some circumstances.

Derived_ Predictions can not be used in conjunction with sensitivities calculated by the process model (SenMethod=-1 or 0 in the Parameter_Data input block). Rather, if the quantity of interest cannot be calculated by the process model, sensitivities for derived predictions need to be calculated by perturbation (SenMethod=1 or 2).

In the following example of the Derived_ Predictions input block, the derived predictions from the example for the Prediction_Data input block are defined here instead. In practice they can only appear in one of the input blocks.

```
#For GroupName=FlowPred, MeasStatFlag=SD in
#the Prediction Groups input block
#Double quotes are needed for the equations because they
#include spaces.
BEGIN Derived Predictions TABLE
  NROW=2  NCOL=5  COLUMNLABELS
  predname refvalue measstatistic equation          groupname
  flow-t3-ss 0.3      0.55      "flow-t3 - flow-ss"  flowpred
  flow-t12-ss .2       0.45      "flow-t12 - flow-ss" flowpred
END Derived_Predictions
```

Chapter 9: INPUT TO INCLUDE MEASUREMENTS OF PARAMETER VALUES

Parameter values can sometimes be directly measured. For example, in ground-water problems pump tests can be used to measure hydraulic conductivity. These measurements can be valuable. Their utility is addressed by Guideline 5 of Table 1 in Chapter 3.

Prior information data are input using the Prior_Information_Groups, which provides data related to groups of prior information on the estimated parameters, and the Linear_Prior_Information block, which provides data for each individual prior item.

Prior_Information_Groups Input Block (optional)

The Prior_Information_Groups input block is used to assign data that apply to many or all of the items of prior information within defined groups. Data for individual items of prior information can be assigned in the Linear_Prior_Information input block, which is described in the next section. The information entered here is replaced by data entered in the Linear_Prior_Information input block.

Keywords this input block include:

GroupName   - Name for a group of prior information items (up to 12 alphanumeric characters including _, i.e. underscore, not case sensitive). Default=DefaultPrior.

UseFlag   -yes: include this group when estimating parameters. no: do not include this group. Default=yes.

PlotSymbol   - An integer used in post-processing programs for the purpose of assigning symbols for plotting. Default=1.

WtMultiplier   - A value that multiplies the weight for each member of the group when the weighting is defined using Statistic and StatFlag keywords described for the Linear_Prior_Information input block. Default=1.0

A CovMatrix is needed to represent correlation between errors in the members of a group. One matrix is used to define the weighting for all the members in a group. In the matrix, members with independent errors will have off-diagonal terms that equal zero.

CovMatrix   - Name of the error variance-covariance matrix. The matrix is specified using the Matrix_Files input block, which is described in Chapter 10. CovMatrix should not be specified if there is no correlation between errors in the members of a group.

Other keywords   - Any keyword from the Linear_Prior_Information input block.

If Blockformat KEYWORDS is selected by designation or default, keywords associated with a prior information group in the Prior_Information_Groups input block need to be grouped together and follow the related GroupName. The GroupName keyword needs to be the first keyword on a new line.

If Blockformat TABLE format is selected, COLUMNLABELS are needed because there is no default column order for the Prior_Information_Groups input block.

Example of the Prior_Information_Groups input block:

```
BEGIN Prior Information Groups
  GroupName=PRIOR  WtMultiplier=1. USEFLAG=yes
  statistic=0.0004 statflag=var plotsymbol=3
END Prior Information Groups
```

Linear_Prior_Information Input Block (optional)

The Linear_Prior_Information input block provides information about individual items of prior information. For each item of prior information, all six of the keywords listed below need to be defined either in the Linear_Prior_Information or the Prior_Information_Groups input block.

By specifying a groupname in this block, all values associated with that group as defined in the Prior_Information_Groups input block are assigned. Keywords defined in the Prior_Information_Groups input block only need to appear here to change the designation.

Items to be specified for each individual prior information item include:

PriorName   - Prior information equation name (up to 20 alphanumeric characters including _, i.e. underscore, not case sensitive). Each name needs to be unique.

Equation   - An equation without an "equal" sign that defines the prior information in terms of parameter names as specified in the Parameter_Data or Derived_Parameters input blocks.

The following rules apply to Equation. Though the same equation capability used by observations and predictions and described in Chapter 13 is used by Linear_Prior_Information, not as many features can be used because the equations need to be linear.

1. The equations can include integer or real numbers and the characters + and *. They can not include the two characters / and ^. They can only include parentheses [the two characters "(" and ")"] in the only allowed function, Log10 (see item 3). Equations can contain spaces only if the equations are enclosed in double or single quotes.

2. Equations can cite any ParamName in the Parameter_Data or Derived_Parameters input blocks.

3. Equations need to be linear. If the parameter is log-transformed it needs to be log-transformed in the equations specified here, using log base 10. Thus, for a parameter named K with Transform=yes in the Parameter_Data or Parameter_Groups input Block, Log10(K) would need to appear in the equation.

PriorInfoValue   - Value of prior information. For log-transformed parameter, specify the native, untransformed value here. UCODE_2005 will log-transform the value.

Statistic   - Value used to calculate the prior information weight.

If the parameter is log-transformed the statistic needs to be related to the base 10 log-transformed parameter. For example, for log-transformed values for which the data specify the value within plus and minus two orders of magnitude with a 95-percent probability, Statistic=1.0 for StatFlag=SD. This is consistent with the following numerical example: PriorInfoValue=0.1; the true value is expected to be between 0.001 and 10.0 (plus and minus two orders of magnitude) with a 95-percent probability. This example suggests that in most circumstances the standard deviation is likely to be less than 1.0. For more information, see Hill (1998, p. 48) and Hill and Tiedeman (in press, chapter 5 and guideline 6).

StatFlag - Character string that defines the corresponding statistic and how it is used to calculate the weight. No default. Options are:

StatFlag	Statistic	Weight calculated as
VAR	Variance	1/Statistic
SD	Standard deviation	$1/(Statistic)^2$
CV	Coefficient of variation	$1/(Statistic \times PriorInfoValue)^2$
WT	Weight	Statistic
SQRWT	Square root of the weight	$Statistic^2$

GroupName - Name for a group of prior information items.

GroupName can be composed of up to 12 alphanumeric characters including _ (underscore), and is not case sensitive. Specifying a GroupName transfers data defined in the Prior_Information_Groups input block for the group to this item. Data specified in the Linear_Prior_Information input block overrides data provided for the group in the Prior_Information_Groups block.

If Blockformat KEYWORDS is selected by designation or default, keywords defining an item of prior information in the Linear_Prior_Information input block need to be grouped together and follow the related PriorName. The PriorName keyword needs to be the first keyword on a new line. PriorName and associated keywords are repeated to define multiple items of prior information.

If blockformat TABLE is selected without indicating ColumnLabels, the default column order is used and is listed below. No columns are ignored and a column for each keyword is expected. If ColumnLabels are indicated, the column labels can appear in any order; the PriorName keyword need not be first, though it often is first.

Default column order: `PriorName Equation PriorInfoValue Statistic StatFlag GroupName`

Example of the Linear_Prior_Information input block (Prior_Information_Groups block is included for completeness):

```
BEGIN PRIOR INFORMATION GROUPS
  GROUPNAME=PRIOR WTMULTIPLIER=1. USEFLAG=YES PLOTSYMBOL=3
  STATFLAG=VAR
END PRIOR INFORMATION GROUPS
```

```
BEGIN Linear Prior Information TABLE
#SS 1 and HK 1 are transformed
#PriorInfoValue is in native space
#Statistic is in log10 space
  nrow=4  ncol=5  columnlabels  GROUPNAME=PRIOR
  PriorName      Equation    PriorInfoValue Plotsymbol Statistic
  Eqn Wells Tr wells tr         -1.0           3         0.1
  Eqn SS 1       log10(SS 1)    1.0e-2         3         1.0
  Eqn HK 1       log10(HK 1)    1.0e-3         3         1.0
  Eqn Rch Ann   .5*Rch1+.5*Rch2 37.0           4         4.0
END Linear Prior Information
```

Chapter 10: INPUT TO DEFINE WEIGHT MATRICES

When the errors in observations are correlated, the errors are described statistically not only using variances, but also covariances. The resulting variance-covariance matrix is inverted to create the weight matrix (Hill and Tiedeman, in press, Chapter 3.1.3; Hill, 1998, p. 13-14). The Matrix_Files input block is used to read the error variance-covariance matrix.

Before deciding to use a full weight matrix, it can be useful to consider using the difference between two measured values. For example, two nearby wells may not have accurately measured elevations, but their elevations relative to one another may be known precisely. In this situation, it may be more useful to use the difference in hydraulic head between the two wells in regression instead of the hydraulic head in each of the wells. Additional examples of using differences in regression are discussed by Hill and Tiedeman (in press) and Hill (1998).

For some situations, correlations can not be avoided. In ground-water studies, one circumstance is when streamflow gain and loss observations are derived from a set of streamflow measurements located sequentially along a single stream with no tributaries, as discussed by Hill (1992), Christensen and others (1998), Hill (1998, p. 47), and Hill and Tiedeman (in press, guideline 6). Consider the situation in which three streamflow measurements, Q1, Q2, and Q3, are used to produce two streamflow gain or loss observations, Q2-Q1 and Q3-Q2. The variance of each of the two streamflow gains or losses equals the sum of the variances of the streamflow measurements subtracted to get the gain or loss. The covariance between adjacent streamflow gain and loss measurements is the negative variance of the shared streamflow measurement.

These relations can be extended directly for a river with more than three sequential streamflow measurements. For example, consider a set of seven streamflow measurements each with a variance of 1.0. The variance-covariance matrix of the six streamflow gain and loss errors equals:

```
 2.   -1.    0.    0.    0.    0.
-1.    2.   -1.    0.    0.    0.
 0.   -1.    2.   -1.    0.    0.
 0.    0.   -1.    2.   -1.    0.
 0.    0.    0.   -1.    2.   -1.
 0.    0.    0.    0.   -1.    2.
```

This example is used in this chapter to illustrate the capabilities of the Matrix_Files input block.

Matrix_Files Input Block (optional)

If the keyword CovMatrix is used to define the names of one or more matrices in the Observation_Groups and(or) Prior_Information_Groups input blocks, a Matrix_Files input block is needed. One matrix needs to be defined for each specified CovMatrix name. In this work the matrices contain the variances (diagonal terms) and covariance (off-diagonal terms) of errors in the observations or prior information.

All observations or prior information with intercorrelated errors need to be included in a single group of the Observation_Groups or Prior_Information_Groups input blocks. Multiple groups of observations or prior information can have intercorrelated errors. For example, errors in observations obs1, obs2, and obs3 may be intercorrelated, and errors in observations obs4, obs5, and obs6 may be intercorrelated. The lack of correlation in errors between the members of the first group and the second group means that these observations can be assigned to two different groups, or they can be assigned to a single group for which only selected covariances are non-zero.

Each column of the variance-covariance matrix is associated with one observation or prior information, and the same order is repeated for the rows of the matrix. The order of otems in the matrix is determined by the order the observations or prior information are defined in the Observation_Data or Linear_Prior_Information input blocks.

There are two keywords of the Matrix_Files input block.

MatrixFile   - Name of or path to the file from which one or more matrices are read. (Up to 2,000 characters; case sensitivity depends on the operating system).

NMatrices   - Number of matrices to be read from MatrixFile. Default=1.

NMatrices can be omitted if the default value is used. If it is included and block format KEYWORDS is used, then keywords MatrixFile and NMatrices need to be listed in pairs and in the order shown.

If Blockformat TABLE format is selected, COLUMNLABELS are needed because there is no default column order for the Matrix_Files input block.

The files listed using MatrixFile can include matrices entered in complete or compressed format. The complete format is useful when the matrix has few zero values. The compressed format is useful when the matrix contains a large number of zero values.

Example of the Matrix_Files input block:

```
BEGIN Matrix_Files KEYWORDS
  MatrixFile=matrix.dat  NMatrices=1
END Matrix_Files
```

Complete Matrix

The input format of a complete matrix is:

```
CompleteMatrix  [NAME]
NGMEM   NGMEM  [ControlRecord]
[Array Control Record]
VAL(1,1)  . . . . . . . VAL(1,NGMEM)
   .      .           .
   .      .           .
   .           .      .
VAL(NGMEM,1) . . . . . . . VAL(NGMEM,NGMEM)
```

where,

CompleteMatrix - A keyword that identifies that this matrix is being read in complete matrix format (case-insensitive).

NGMEM - The number of members in the group being considered. The matrix dimensions are (NGMEM, NGMEM). The value of NGMEM needs to equal the number of observations or prior information defined for the group in the Observation_Data or Linear_Prior_Information input block.

ControlRecord - Optional keyword which, when present, results in the array being read as defined by an optional array control record.

Array Control Record - Optional line that can be used to define how the array is read. The default is to read the array as free format. Construction of the Array Control Record is discussed below.

VAL(i,j) - The value of the matrix element in row i and column j.

The Array Control Record generally is used only if the matrix is to be read from another file or if the matrix can not be read using free format. Free format requires that the numbers be separated by one or more spaces, a comma, a comma and one or more spaces, or are on a new line. Construction of the array control record is discussed later in this chapter.

Example complete matrix using three methods for data input are shown. To be consistent with the example presented for the Matrix_Files input block, the files that start with "CompleteMatrix" would be called "matrix.dat" and located in the same directory. If located in another directory a pathname would need to be specified for Matrix_Files input block keyword MatrixFile.
1. Use the default free format to read the matrix.

```
CompleteMatrix CovPRIOR
6 6
    2.    -1.     0.     0.     0.     0.
   -1.     2.    -1.     0.     0.     0.
    0.    -1.     2.    -1.     0.     0.
    0.     0.    -1.     2.    -1.     0.
    0.     0.     0.    -1.     2.    -1.
    0.     0.     0.     0.    -1.     2.
```

2. Use the Array Control Record to read from a separate file.

```
CompleteMatrix CovPRIOR
6 6  ControlRecord
OPEN/CLOSE    covprior.txt    1.    "(FREE)"   1
```

where file covprior.txt would contain:

```
 2.    -1.    0.    0.     0.    0.
-1.     2.   -1.    0.     0.    0.
 0.    -1.    2.   -1.     0.    0.
 0.     0.   -1.    2.    -1.    0.
 0.     0.    0.   -1.     2.   -1.
 0.     0.    0.    0.    -1.    2.
```

3. Use an Array Control Record with an F11.0 format that requires the numbers to be right-justified in fields that are six places wide:

```
CompleteMatrix CovPRIOR
6 6 ControlRecord
internal  1.0  (3F6.0)  21
 2.    -1.    0.    0.     0.    0.
-1.     2.   -1.    0.     0.    0.
 0.    -1.    2.   -1.     0.    0.
 0.     0.   -1.    2.    -1.    0.
 0.     0.    0.   -1.     2.   -1.
 0.     0.    0.    0.    -1.    2.
```

Compressed Matrix

The input format of a compressed matrix is:

```
CompressedMatrix  [NAME]
NNZ   NGMEM   NGMEM [ControlRecord]
[Array Control Record]
IPOS(1)    VAL(1)
IPOS(2)    VAL(2)
...
IPOS(NNZ)  VAL(NNZ)
```

CompressedMatrix - A keyword that identifies that this matrix is being read in compressed format (case-insensitive),

NNZ - The number of non-zero values in the matrix,

NGMEM - The number of members in the group. The matrix dimensions are (NGMEM, NGMEM).

The value of NGMEM needs to equal the number of observations or prior information defined for the group in the Observation_Data or Linear_Prior_Information input block. NGMEM is repeated because the software can read asymmetric matrices.

IPOS(i) - The position of the ith non-zero entry in the matrix, assuming column-major storage order as described below.

VAL(i) - The corresponding non-zero value.

Array Control Record - Optional line that can be used to define how the array is read. The default is to read the array as free format. Construction of the Array Control Record is discussed below.

In column-major storage order, all entries of column 1 are numbered first, starting at row 1, followed by all entries of column 2, and so on. For example, for a matrix of 6 rows and 6 columns, the column-major storage ordering is as follows:

1	7	13	19	25	31
2	8	14	20	26	32
3	9	15	21	27	33
4	10	16	22	28	34
5	11	17	23	29	35
6	12	18	24	30	36

For the example described in the beginning of this chapter, the matrix contains 16 non-zero values, so NNZ = 16. This matrix would be represented in compressed form as:

```
CompressedMatrix
     16      6      6
     1             1.
     2            -1.
     7            -1.
     8             2.
     9            -1.
    14            -1.
    15             2.
    16            -1.
    21            -1.
    22             2.
    23            -1.
    28            -1.
    29             2.
    30            -1.
    35            -1.
    36             2.
```

Array Control Records

Array control records are read in free format using convenient text-based conventions.

Array Control Record Input Instructions

Entries in bold italics are keywords that can be uppercase or lowercase. Three keywords are possible: CONSTANT, INTERNAL and OPEN/CLOSE.

1. ***CONSTANT*** CNSTNT

   All elements of the array are set to the value supplied by CNSTNT.

2. ***INTERNAL*** CNSTNT FMTIN IPRN

   The array is read from the file that contains the Array Control Record.

3. ***OPEN/CLOSE*** FNAME CNSTNT FMTIN IPRN

   The array is read from the file with a name specified by FNAME. This file is opened just prior to reading the array and closed immediately after the array is read. A file that is read using this Array Control Record can contain only a single array.

Each Array Control Record is limited to a length of 2000 characters.

Explanation of Variables in the Array Control Records

CNSTNT - a real number. For keywords INTERNAL and OPEN/CLOSE , all elements in the matrix are multiplied by CNSTNT after they are read.

FMTIN - the format for reading array elements. The format string needs to contain 100 characters or less. The format string needs to either be a standard Fortran format that is enclosed in parentheses, or "(FREE)", including the double quotes, which indicates free format. When the "(FREE)" option is used, be sure that all array elements have a non-blank value and that a comma or at least one blank separates adjacent values.

IPRN - a flag that indicates whether to print the array to the main output file and determines the format used. Matrices read as compressed are printed as full matrices. If IPRN is less than zero, the array is not printed. IPRN is set to zero when the specified value exceeds those defined. The defined values of IPRN and how two numbers are printed using the different formats are presented after the variable definitions.

FNAME – a file name or an absolute or relative pathname less than or equal to 2,000 characters in length.

IPRN	FORMAT	How the number -3.44234 is printed	How the number -3.44234×10^{-4} is printed
	Column number	123456789012345	123456789012345
0	10G11.4	-3.442	-0.3442E-03
1	11G10.3	-3.44	-0.344E-03
2	9G13.6	-3.44234	-0.344234E-03
3	15F7.1	-3.4	-0.0
4	15F7.2	-3.44	-0.00
5	15F7.3	-3.442	-0.000
6	15F7.4	-3.4453	-0.0003
7	20F5.0	-3.	-0.
8	20F5.1	-3.4	-0.0
9	20F5.2	-3.44	-0.00
10	20F5.3	*****	-.000
11	20F5.4	*****	*****
	Column number	123456789012345	123456789012345
12	10G11.4	-3.442	-0.3442E-04
13	10F6.0	-3.	0.
14	10F6.1	-3.4	-0.0
15	10F6.2	-3.44	-0.00
16	10F6.3	-3.442	-0.000
17	10F6.4	******	-.0003
18	10F6.5	******	******
19	5G12.5	-3.4423	-0.34423E-04
20	6G11.4	-3.442	-0.3442E-04
21	7G9.2	-3.4	-0.34E-04
	Column number	123456789012345	123456789012345

*Number does not fit in the field provided. For many compilers this results in the printing of a series of asterisks, as shown. Positive numbers would be printable in some of the situations for which asterisks are displayed here.

Examples of Array Control Records

The following examples read an array consisting of 4 rows with 7 columns per row:

```
INTERNAL   1.0  (7F4.0)  3      This reads the array values from
 1.2 3.7 9.3 4.2 2.2 9.9 1.0    the file that contains the array
 3.3 4.9 7.3 7.5 8.2 8.7 6.6    control record. The values
 4.5 5.7 2.2 1.1 1.7 6.7 6.9    immediately follow the array
 7.4 3.5 7.8 8.5 7.4 6.8 8.8    control record.
```

```
OPEN/CLOSE  test.dat  1.0  (7F4.0)  3   This reads the array
                                        from the file named
                                        "test.dat". Test.dat
                                        contains only the
                                        array.
```

Chapter 11: INPUT TO INTERACT WITH THE PROCESS MODEL INPUT AND OUTPUT FILES

UCODE_2005 interacts with process model(s) through model input and output files. This chapter describes the tools available for the needed interactions using the Model_Input_Files input block and associated Template files and the Model_Output_Files input block and associated Instruction files.

For UCODE_2005 to run correctly, at least one process-model input file needs to be manipulated via a Template file as defined by the Model_Input_Files input block and one output file needs to be perused to extract simulated equivalents as defined by the Model_Output_Files input block, so both these input blocks are required.

Construct Process-Model Input Files Using Current Parameter Values

When UCODE_2005 runs a model, it first uses data from the Model_Input_Files input block to create model input files from Template input files. There is one template file for each model input file. If many model input files need to be changed, one template file is needed for each of them. The model input files created by UCODE_2005 reflect the most recent parameter values.

Model_Input_Files Input Block (required)

The Model_Input_Files input block lists each process-model input file that needs to be changed when parameter values change and an associated template file. The two items of the Model_Input_Files input block need to occur in pairs. If blockformat KEYWORDS is used by default or designation, MODINFILE needs to be the first keyword and needs to start a new line.

Keywords for the "Model_Input_Files" block include:

ModInFile   - Name for a process-model input file. (Up to 2,000 characters) Names with spaces need to be enclosed in double quotes.

TemplateFile   - Name of the template file that UCODE_2005 will use to create the associated process-model input file, MODINFILE. (Up to 2,000 characters. Case sensitivity depends on the operating system) Names with spaces need to be enclosed in double quotes.

Numerous pairs of MODINFILE and TEMPLATEFILE may be needed depending on the number of files associated with the process model(s) input that are affected by parameters to be adjusted by UCODE_2005.

Default column order: `ModInFile TemplateFile`

Three equivalent example input blocks are shown.

Use the blockformat KEYWORDS:
```
BEGIN MODEL INPUT FILES Keywords
modinfile=..\data Efunc\tc1-fwd.sen templatefile=tc1sen.tpl
modinfile=..\data Efunc\tc1-fwd.mlt templatefile=tc1-fwd.mlt.tpl
END MODEL INPUT FILES
```

Use blockformat TABLE and `COLUMNLABELS` is specified:
```
BEGIN MODEL INPUT FILES TABLE
  NROW=2 NCOL=2 ColumnLabels
  modinfile                templatefile
  ..\data Efunc\tc1-fwd.sen  tc1sen.tpl
  ..\data Efunc\tc1-fwd.mlt  tc1-fwd.mlt.tpl
END MODEL INPUT FILES
```

Use blockformat TABLE and the default column order:
```
BEGIN MODEL INPUT FILES TABLE
  NROW=2 NCOL=2
  ..\data Efunc\tc1-fwd.sen  tc1sen.tpl
  ..\data Efunc\tc1-fwd.mlt  tc1-fwd.mlt.tpl
END MODEL INPUT FILES
```

Template Files

A template file is derived from a model input file, as described in the following sections.

Construction

Consider the model input file shown in Figure 2. On the second line, hydraulic conductivity (K) is defined to equal 0.01, storativity (S) is defined to equal 0.005, and thickness (b) is defined to equal 10. For UCODE_2005 to control K, S, and b, a template file such as that shown in Figure 3 is needed. As discussed in the following section, this file allows UCODE_2005 a large number of spaces within which to substitute values, and this is important in many circumstances.

```
# numberdrawdowns   Q        K       S         b
10 0.75 0.01    0.005     10.
    20.      200.
    20.      400.
    20.      800.
    50.      200.
    50.      400.
    50.      800.
    90.      800.
    90.     1200.
   150.      800.
   150.     1200.
```

Figure 2. A model input file. The values to be represented by parameters are shaded.

```
jtf @
# numberdrawdowns   Q        K       S        b
10 0.75 @K            @   @S              @   @b            @
    20.      200.
    20.      400.
    20.      800.
    50.      200.
    50.      400.
    50.      800.
    90.      800.
    90.     1200.
   150.      800.
   150.     1200.
```

Figure 3. A template file corresponding to the model input file of Figure 2.

Substitution Delimiter

As Figure 3 shows, a template file is created from a model input file by first inserting a line at the top. The line contains "jtf" followed by one or more spaces and the substitution delimiter. jtf stands for JUPITER template file. Choose the substitution delimiter carefully. It can not be any of the characters [a-z], [A-Z] and [0-9], and can not appear in the template file except in its capacity as substitution delimiter. Commonly used substitution delimiters are @ and !.

The substitution delimiter is used to define the field within which UCODE_2005 places a number, as in line 3 of Figure 3. A field that is too small limits the number of significant digits. This, in turn, can affect the accuracy of process model results, sensitivities, regression results, and so on. If the field is so small that the number can not be printed at all, execution stops and an error message is printed.

The substitution field is between and includes a pair of substitution delimiters. All of the characters between and including the substitution delimiters are replaced. Characters between the delimiters need to include one ParamName defined in the Parameter_Data or Derived_Parameter input block. The ParamName can be placed anywhere between the delimiters. All other characters between the delimiters need to be spaces.

The width available for substitution depends on the requirements of the process-model input files. Generally models read numbers: (1) from fields of a specified width or (2) as a sequence of numbers each of which may be of any width and need to be separated by at least one space, a comma, or other character.

In circumstance 1, the substitution widths are prescribed by the process model input instructions. In this case there need not be space or other character between substitution fields. So, for example, the following sequence could occur:

```
@K        @@S          @
```

where 10 spaces are allocated for each number by the process model.

Circumstance 2 is sometimes referred to as free format. In this circumstance, the substitution fields generally can be defined as large as desired, but they need to be separated by a space or a comma so the program knows where one number ends and the next number begins. The input file shown in Figure 2 and Figure 3 has free format specified for the substituted numbers. As a result, the widths for the numbers can be different in the two figures and spaces are placed between the substitutions fields in Figure 3.

Read from Process-Model Output Files

After running the process model(s), UCODE_2005 extracts (reads) values from the output files. These values are used either directly or indirectly to produce simulated equivalents that are compared to observed values or to produce predictions. The values are said to be used indirectly for derived observations and derived predictions, where derived applies to those defined using an equation (see Chapter 8).

Model_Output_Files Input Block (required)

The Model_Output_Files input block defines how UCODE_2005 obtains values from the files produced by the process model. The three keywords of the Model_Output_Files input block need to occur as a set, and the set needs to be repeated for each process-model output file read. If block format is defined by default of designation as KEYWORDS, the ModOutFile keyword needs to be listed first for each set.

ModOutFile — Name of the process-model output file from which UCODE_2005 is to extract values. ModOutFile can be up to 2,000 characters; case sensitivity depends on the operating system. Names with spaces need to be enclosed in double quotes.

InstructionFile — Name for the Instruction file that UCODE_2005 uses to extract values from ModOutFile. InstructionFile can be up to 2,000 characters; case sensitivity depends on the operating system. Names with spaces need to be enclosed in double quotes.

Instruction file construction is described in the following sections. Instruction files can not contain any comment lines or blank lines, even at the end of the file.

ModOutFile and InstructionFile can be file names, in which case the files need to be located in the directory where UCODE_2005 is executed, or they can be an absolute or relative pathnames.

Category — The following options are available:

Obs – The process-model output file is used to calculate simulated equivalents to observations (These files are ignored when prediction=yes in the UCODE_Control_Data input block)

Pred –The process-model output file is used to calculate predictions (These files are used only for modes that use predictions, and otherwise ignored)

Default column order: `ModOutFile InstructionFile Category`

Example using Blocklabel Keywords:

```
BEGIN MODEL OUTPUT FILES Keywords
  modoutfile=tc1-fwd. os  instructionfile=tc1.ins   category=obs
END MODEL OUTPUT FILES
```

or, using block format TABLE to define two files:

```
BEGIN MODEL OUTPUT FILES TABLE
  nrow=2 ncol=3 columnlabels
  modoutfile          instructionfile        category
  tc1-fwd. os         tc1-fwd. os.ins          obs
  tc1-fwd.lst         tc1-fwd.lst.ins          obs
END MODEL OUTPUT FILES
```

Instruction Input File for a Standard Process-Model Output File

The complexity of the instruction file depends on the structure of the process-model output file. An output file from which one column of numbers is read is called a standard file. Standard files need a simple instruction file described in this section.

A standard file can have lines at the top that are to be skipped (the number of lines can be zero). Subsequently, the data need to be in columns separated by spaces or commas. Simulated values are read from each line of data, and always from the same column.

Standard files can be read using an instruction file with the following components.

```
jif @
StandardFile Nskip ReadColumn Nread
Names for each of the Nread values.
Place each name on a new line.
```

where

jif @   defines the file delimiter (see below) needed in all instruction files. It is not used to read standard files.

StandardFile is a keyword that indicates a standard file is being read.

Nskip is the number of lines to skip at the beginning of the file and can equal 0 or any positive integer.

ReadColumn is the column of the file from which values are to be read.

Nread is the number of values, and therefore lines, to be read.

The "Names for each of the Nread values" listed on the subsequent lines need to match observation names specified in the Observation_Data or Derived_Observations input blocks if Category=Obs. They need to match prediction names specified in the Prediction_Data or Derived_Predictions input blocks if Category=Pred. Category is defined for each instruction file in the Model_Output_Files input block.

The entries on the line with keyword StandardFile are read in free format, so they need to be separated by a space or a comma, or end of line. Instruction files are not allowed to contain any comment lines or blank lines, even at the end of the file.

Execution time can be reduced if the names in the process-model output file are in the same order as they are defined in the Observation or Prediction input blocks. If the order is unclear, execute the process model(s) and look at the files to be read.

For example, five simulated values can be read as a standard file from an _os output file produced by MODFLOW_2000. A MODFLOW_2000 _os file and an associated

instruction file are shown below. The instruction file skips zero lines and reads five
values from the first column of the _os file:

First five lines of file tc1._os from the MODFLOW 2000 distribution[1]				
100.1747	101.8040	1	1.0	0.000000
-0.9155273E-04	-0.2899933E-01	1	1.1	87163.00
-0.8264923E-01	-0.1289978	1	1.12	0.2443906E+08
126.9926	128.1170	1	2.0	0.000000
-0.3369141E-01	-0.4100037E-01	1	2.1	87163.00
Instruction file				
jif @ StandardFile 0 1 5 Obs1 Obs2 Obs3 Obs4 Obs5				

[1] The _os file produced by MODFLOW_2000 version 1.15 has no header line and has a
fifth column of real numbers that equal the observation time. An _os data exchange file
produced by UCODE_2005 has one header line with column labels and no fifth column
of real numbers.

For observations, the numbers read are compared with five observations defined in the
Observation_Data or Derived_Observations input block corresponding to the names
listed in this instruction file.

Another example shown in Figure 4 includes the output file written by the model whose
input file appears in Figure 2. Suppose that parameters in the template file of Figure 3 are
estimated by minimizing the squared differences between the drawdowns calculated by
the model (Figure 4) and those measured in the field. The simulated drawdown values
can be read using the StandardFile instruction file:

```
jif @
StandardFile 7 5 10
dd1
dd2
dd3
dd4
dd5
dd6
dd7
dd8
dd9
dd10
```

```
NUMBER OF OBSERVATIONS.......... =       10.
 PUMPING Q..................... = 7.50000E-01
 K.AQUIFER..................... = 1.00000E-03
 STORAGE COEFFICIENT.AQUIFER .... = 5.00000E-03
 THICKNESS of AQUIFER........... = 1.00000E+01
 TRANSMISSIVITY................. = 1.00000E-02
    RAD       TIME          u             W           DRAWDOWN
   20.00      200.00    0.2499999881   1.0442823072    6.2325862595
   20.00      400.00    0.1249999941   1.6234253182    9.6890833659
   20.00      800.00    0.0624999970   2.2569095812   13.4699051666
   50.00      200.00    1.5624999258   0.0911875383    0.5442342502
   50.00      400.00    0.7812499629   0.3213536892    1.9179340435
   50.00      800.00    0.3906249814   0.7183524996    4.2873405859
   90.00      800.00    1.2656249399   0.1428815143    0.8527592170
   90.00     1200.00    0.8437499599   0.2871831829    1.7139943357
  150.00      800.00    3.5156248330   0.0068363119    0.0408011353
  150.00     1200.00    2.3437498887   0.0306533241    0.1829481216
```

Figure 4. An example model output file in which the numbers of interest are shaded.

Instruction Input File for a Non-Standard Process-Model Output File

More complex situations require more complex capabilities, and UCODE_2005 provides a comprehensive set of instruction options for non-standard output files. These instructions are identical to those provided in PEST (Doherty, 2004) due to Doherty's involvement in development of the JUPITER API (Appendix A). The discussion below is derived from the discussion in the JUPITER API documentation, which is largely drawn from the PEST documentation.

UCODE_2005 provides comprehensive instructions that permit a value in a model output file to be found in the same way that a person would look for a number. A person would skim down the file looking for something recognizable - a "marker" - that is near a value of interest and place their cursor at that location, ready to do the next task. For example, if a value of interest is the calculated flow after the first time step in the third stress period, an instruction could identify a marker such as:

```
VOLUMETRIC BUDGET FOR ENTIRE MODEL AT END OF TIME STEP 1 STRESS PERIOD 3
```

The value of interest may be located, for example, on the 4th line following the marker between character columns 23 and 30; or on the 3rd line after the marker, the 5th item from the left; and so on. Markers can be primary or secondary. Primary markers are used to find a line in the file. Secondary markers are used to navigate within a line.

Instruction files are not allowed to contain any comment lines or blank lines, even at the end of the file.

An Example Instruction File for a Non-Standard Input File

If the example shown in Figure 4 were read without the StandardFile option, the instruction file required is shown in Figure 5. The details of this file as constructed are described in the following sections of this report.

```
jif @
@ RAD      TIME @
11 [dd1] 53:69
11 [dd2] 53:69
11 [dd3] 53:69
11 [dd4] 53:69
11 [dd5] 53:69
11 [dd6] 53:69
11 [dd7] 53:69
11 [dd8] 53:69
11 [dd9] 53:69
11 [dd10] 53:69
```

Figure 5. A non-standard instruction file that reads the shaded numbers in Figure 4.

Preliminaries

Three things need to be defined to begin describing non-standard input files: the marker delimiter, extraction names, and extraction type.

Marker Delimiter, jif

The first line of an instruction file needs to begin with the three letters "jif" (stands for JUPITER instruction file), a single space, and a single character. The character is used as the marker delimiter, which defines the extent of a marker. A marker could also be called a search string; it is a sequence of characters that is the object of a search. Markers need to fit on one line of the instruction file. A marker delimiter immediately precedes the first character of a marker and immediately follows the last character. For example, given "%" as the delimiter, %LAYER 1% identifies the marker "LAYER 1".

A marker delimiter can not be any of the following characters because these characters have other roles in instruction files.
A B C D E F... Z
abcdef ... z
0 - 9
!
[
]
(
)
:
&

The character chosen for a marker delimiter also cannot occur within the text of any markers.

Usually \$, @, % or ~ are good choices for a marker delimiter.

Extraction Names

The name used for an extracted value may be the name of an observation or a prediction or may be 'dum', which indicates that the value is not used. As illustrated below, 'dum' can be useful in navigating a model output file.

Except for 'dum', a different name needs to be used for each extracted value. This applies to all the values extracted using all the instruction files defined for one UCODE_2005 file set.

The Instruction Set

The instructions are listed in Table 9 and are described in detail in the following section. In Table 9, the instructions are divided into those that need to start at the beginning of the line and those that need to start later on the line.

UCODE_2005 uses the instructions to read a model output file from the top toward the bottom. Instructions cannot move backward to a previous line of the model output file or backward on a line. If there is more than one value to be extracted from one line of the model output file, they need to be extracted from left to right.

Table 9. Instructions available in UCODE_2005.
[All instruction lines need to start with something in the first column. Each instruction needs to be bounded by the beginning of a line, a blank space, or the end of a line. No blank lines or comments are allowed in instructions files. #, any positive integer; name, 'dum' or an observation or a prediction name; a, b, column numbers. Column numbers are obtained by counting from the left side of a line, starting at 1.]

Instruction name	Format	Description	Example
Instructions that need to begin in the first column in a line of an instruction file			
Primary marker	Marker delimiter in the first column of a line in an instruction file, followed by a search string, followed by a marker delimiter.	Identifies a search string used to find a line within the process-model output file. The entire search string needs to be on one line.	%find me%
Line advance	l#, where l is a lower case L	Advances # lines down from the present line. When a file is opened the line is 0. Use l1 (line 1) to reach the first line.	l5
Continuation	& positioned in the first place in a line of an instruction file. Markers can not be broken across lines.	The instruction continues from the previous line.	&
Instructions that need to begin later in a line of an instruction file			
Secondary marker	Marker delimiter not in the first column of an instruction file, followed by a search string, and ending with a marker delimiter.	Identifies a search string to navigate within a line of the process-model input file. Secondary markers are preceded on the instruction file line by one of the instructions above.	l5 %find me%
Whitespace	w	Advance in a line to the last blank in the next set of blanks. Repeat as needed.	l5 w w
Tab	t#	Advance in a line to column #.	l5 t60
Extractions			
Fixed	[name]a:b	Read a number from a line starting at column a and ending at column b.	[dd1]1:10
Semi-fixed	(name)a:b	Read a number that may be only partially con-tained in the range extend-ing from a to b, inclusive.	(dd1)2:8
Non-fixed	!name!	Read a number in free format. The number needs to be bounded by spaces, the beginning of a line, or the end of a line.	!dd1!

Extraction Type

When a number is read from a file it is said to be extracted. Extractions can be categorized into three types: fixed, semi-fixed, and non-fixed (Table 9). They are discussed briefly here and in more detail in the section below entitled "Example instruction files".

Fixed extraction instructions consist of two parts. The first part consists of the extraction name enclosed in square brackets; the second part consists of the range of columns (inclusive) from which to extract the value. These parts cannot be separated by a space because a space always indicates the start of a new instruction unless the space lies within a marker.

Semi-fixed extraction instructions are similar to the instructions for fixed extractions except that the extraction name is enclosed in parentheses instead of square brackets. In semi-fixed extractions, the column numbers do not need to exactly locate the number to be read. Part of the number may precede location a or continue past location b. At least part of the number needs to be located between locations a and b, and the interval can not include digits from any other number. In the output file, each end of the extracted number needs to be bounded by a space, the beginning of a line, or the end of a line.

A semi-fixed instruction operates as follows. It sends the cursor to the first of the two listed columns. If this column is occupied by a space, it searches to the right until it reaches either a non-space, which is taken to be the first character to be extracted, or the second listed column. If it reaches the second column without finding a non-space, an error condition arises. If it finds a non-space, it then searches to the right for the next blank or the end of the line, either of which would be taken to bound the right side of the extraction. If the column identified by the first number is non-blank, the first character of the extraction is sought to the left and the last character of the extraction is sought to the right. The width of a number read through a semi-fixed abstraction can be longer than the difference between the column numbers cited in the semi-fixed extraction instruction.

Non-fixed extraction instructions consist of the extraction name enclosed by exclamation marks. The number is read using free format; methods other than column numbers are used to locate the number within a line of the process-model output file. A value to be extracted by non-fixed extraction needs to be bound by spaces, the beginning of a line, or the end of a line.

Primary Marker

Primary markers are used to find a line of an output file that contains a specific string of characters. If a marker is the first item on an instruction line, then it is a primary marker; if it occurs later in the line, following other instruction items, it is a secondary marker.

On encountering a primary marker UCODE_2005 reads the model output file, line by line, searching for the string between the marker delimiter characters. When it finds the string it positions itself at the last character of the string. Any further instructions on the instruction line are used to process the remainder of this line of the process-model output file. If there are spaces in a primary (or secondary) marker, exactly the same number of spaces is expected in the matching string on the model output file.

Often, a primary marker is part or all of a header or label because it precedes a list of results and thus makes a convenient reference point.

Searching for a primary marker is time-consuming. If the values to be read are always written to the same lines of a model output file for every model run, the line advance instruction can be used and is more efficient. If a primary marker is needed, execution times are faster if the marker has as many characters as possible.

Use of primary markers is shown in Figure 6. Figure 6 first shown parts of the process-model output file and then shows two instruction files paired with the line to be read in which the digits identified by the extractions are shaded. The instruction files shown in Figure 6 extract the numbers comprising the third solution vector. The primary marker "TIME PERIOD NO. 3 --->" establishes a reference point for searching for "SOLUTION VECTOR:"; if this reference point were not established a solution vector from an earlier time would be read.

Line Advance, l#

The syntax for the line advance item is "l#" where # is the number of lines to advance; note that "l" is a lower case "L", not "one". The line advance item needs to start in column 1 of an instruction line (table 9). In figure 6 model-calculated solution vectors are written on the line after the marker, so the cursor is instructed to move to the beginning of a new line using the "l1" (line 1) line advance item.

The cursor starts on line zero. Thus if the first instruction line begins with "l1" (line1), processing of the model output file begins on its first line; similarly, if the first instruction line begins with "l8", processing of the model output file begins on its 8th line.

Portion of model output file. The numbers to be extracted are shaded.	``` . . TIME PERIOD NO. 1 --->`         .         . SOLUTION VECTOR:   1.43253   6.43235   7.44532   4.23443   91.3425         .         . TIME PERIOD NO. 2 --->` . . SOLUTION VECTOR: 1.34356 7.59892 8.54195 5.32094 80.9443 . . TIME PERIOD NO. 3 --->` . . SOLUTION VECTOR: 2.09485 8.49021 9.39382 6.39920 79.9482 ```

Instruction file 1 using semi-fixed extractions	```
	jif *
	        .
	        .
	TIME PERIOD NO. 3 --->
	SOLUTION VECTOR:
	l1 (obs1)5:10 (obs2)12:17 (obs3)21:28 (obs4)32:34
	& (obs5)41:45
	``

Columns defined by instruction file 1 are shaded	```
	  2.09485 8.49021 9.39382 6.39920 79.9482
	1234567890123456789012345678901234567890123456789
	         10 20 30 40
	``

Instruction file 2 using fixed extractions	```
	jif *
	        .
	        .
	TIME PERIOD NO. 3 --->
	SOLUTION VECTOR:
	l1 [obs1]3:9 [obs2]12:18 [obs3]21:27 [obs4]30:36
	& [obs5]39:45
	``

Columns defined by instruction file 2 are shaded	```
	  2.09485 8.49021 9.39382 6.39920 79.9482
	1234567890123456789012345678901234567890123456789
	         10 20 30 40
	``

Figure 6. Portion of a model output file illustrating the use of multiple primary markers, line advance, and semi-fixed reading. The numbers being read are shaded and column numbers are shown. Dots replace one or more lines that are not shown. Instruction file 1 uses semi-fixed reading; instruction file 2 uses fixed reading. Both instruction files extract the same values.

Continuation, &

An instruction line can be broken between any two instructions by using the continuation character, "&", to indicate that an instruction line is actually a continuation of the previous line. Thus the instruction file line

```
11 %RESULTS% %TIME (4)% %=% !obs1! !obs2! !obs3!
```

is equivalent to

```
11
& %RESULTS%
& %TIME (4)%
& %=%
& !obs1!
& !obs2!
& !obs3!
```

For both, the marker delimiter is assumed to be "%". Note that the continuation character needs to be followed by at least one space before the next instruction.

Secondary Marker

A secondary marker instructs the cursor to move along the current model output file line until it finds the secondary marker string, and to place the cursor on the last character of that string.

The first part of figure 7 shows a portion of a model output file while the second part of figure 7 shows instructions for reading the potassium (K) concentration. A primary marker is used to place the cursor on the line above the line where the calculated concentrations are recorded for the distance of interest. The next instruction advances one line and reads the number following secondary marker "K:" to extract a value named "kc".

A useful feature of the secondary marker is illustrated in Figure 8. If a secondary marker is preceded only by a primary marker, the primary and secondary markers need to be found on the same line. Thus the instruction "%TIME STEP 10%" causes the cursor to pause on its downward journey through the model output file at the first line illustrated in the top part of Figure 8. However, when it does not find the string "STRAIN" on the same line it continues looking for the string "TIME STEP 10" again. Eventually it finds a line containing both the primary and secondary markers and then begins execution of the next instruction.

Multiple secondary markers can be specified. If they can be combined into one long primary marker, execution time can be reduced.

If a secondary marker is unmatched and the first instruction on the line is not a primary marker, execution is terminated with an appropriate error message. As mentioned above, if the first instruction is a primary marker, the search continues until a match is found or the end of the file is reached.

Portion of model output file	. . DISTANCE = 20.0: CATION CONCENTRATIONS:- Na: 3.49868E-2 Mg: 5.987638E-2 K: 9.987362E-3 . .
Portion of instruction file	jif ~ . . ~DISTANCE = 20.0~ ll ~K:~ !kc! . .

Figure 7. Portion of a model output file and related lines in the instruction file illustrating ll (line 1) preceding a secondary marker. The number being read is shaded. Dots replace one or more lines that are not shown.

Portion of model output file	. . TIME STEP 10 (13 ITERATIONS REQUIRED) STRESS ---> X = 1.05 STRESS = 4.35678E+03 X = 1.10 STRESS = 4.39532E+03 . . TIME STEP 10 (BACK SUBSTITUTION) STRAIN ---> X = 1.05 STRAIN = 2.56785E-03 X = 1.10 STRAIN = 2.34564E-03 . .
Portion of instruction file	jif % . %TIME STEP 10% %STRAIN% ll %STRAIN =% !str2! . .

Figure 8. Portion of a model output file and related lines in the instruction file illustrating use of a primary marker and secondary marker on one instruction line. The numbers being read are shaded. Dots replace one or more lines that are not shown.

Whitespace, w

The whitespace instruction directs the cursor to move forward from its current position until it encounters the next blank space, and continue until it finds a non-blank space, finally stopping on the last blank space in the sequence of blank spaces. The whitespace instruction is a "w", separated from neighboring instructions by at least one blank space.

Consider the model output file line presented in Figure 9. The instruction line begins with a primary marker. After this marker is found and processed the cursor rests on the ":", which is the last character of the marker string. In response to the first whitespace instruction the cursor finds the next blank space and then moves to the last of this series of blank spaces; that is, just before the "2" of the first number on the line. The second whitespace instruction moves the cursor to the space preceding the first "4" of the second number on the line; processing of the third whitespace instruction results in the cursor moving to the space just before the negative sign. After the fourth whitespace instruction is implemented, the cursor rests on the space preceding the last number; the latter is then read as a non-fixed dependent.

Portion of model output file	. . . MODEL OUTPUTS: 2.89988 4.487892 -4.59098 8.394843 . .
Instruction file	jif % %MODEL OUTPUTS:% w w w w !obs1!

Figure 9. Portion of a model output file and instruction file illustrating use of the white space instruction. The number being read is shaded. Dots replace one or more lines that are not shown.

Tab, tn

The tab instruction places the cursor after a user-specified column on the model output file line that it is currently processing. The instruction syntax is "t#" where # is a column number. Like the whitespace instruction, the tab instruction can be used to locate a non-fixed extraction. Use of the tab instruction is illustrated in Figure 10.

Portion of model output file	. . . TIME(1): A = 1.34564E-04, TIME(2): A = 1.45654E-04
Column numbers	12345678901234567890123456789012345678901234567890 10 20 30 40 50
Instruction file	jif % 14 t34 %=% !a2!

Figure 10. Portion of a model output file and instruction file illustrating use of the tab instruction. The number read is shaded. Dots replace one or more lines that are not shown.

The instructions shown in Figure 10 assume that the cursor was previously four lines above the line shown; the marker delimiter is defined to be "%". Implementation of the "t34" instruction places the cursor after the ":" following the "TIME(2)" string. The secondary search string then moves the cursor to the next "=" character. From there it reads the next number on the line (in this example, it is the last number on the line) as a non-fixed dependent.

Example Instruction Files

The examples are organized based on their use of fixed, semi-fixed, and non-fixed extractions.

Fixed Reading

The second instruction file of Figure 6 illustrates the use of fixed reading and shows how the same numbers can be read using semi-fixed reading. Another example is shown in Figure 11.

Line from model output file	1236.567 8495.000 -900.000
Column numbers	12345678901234567890123456789012345678901234567890 10 20 30 40 50
Instruction line	11 [A]1:8 [B]10:17 [C]19:26

Figure 11. Portion of an instruction line illustrating fixed reading.

Fixed reading is useful when the model writes its output in tabular form using fixed field-widths. The range specified needs to be wide enough to accommodate the maximum length that the number will occupy in the course of all model runs. If it is not wide enough, the extracted number may be truncated or a negative sign may be omitted. However, the range must not be so wide that it includes part of another number; in this case an error occurs, and execution stops after an error message is printed.

For model results that are written as an array of numbers, the numbers may be immediately adjacent to one another so that there are no intervening spaces and fixed reading is required. For example, the numbers shown in Figure 12 require fixed reading. Semi-fixed and non-fixed reading require intervening spaces or an always present non-numeric string that can be used as a marker.

Line from model output file	1236.5678495.000-900.000
Column numbers	12345678901234567890123456789012345678901234567890 10 20 30 40 50
Instruction line	1 [A]1:8 [B]9:16 [C]17:24

Figure 12. Portion of a model output file and instruction file illustrating use of fixed reading. The three numbers being read are shaded; each has three digits to the right of the decimal place. Dots replace one or more lines that are not shown.

Semi-Fixed Reading

Figure 6 demonstrates the use of semi-fixed reading.

Semi-fixed reading is useful if the placement of a number always occupies some spaces, but the rest of the range depends on the magnitude of the number. The only requirement is that part of the number will always fall between (and including) the two listed columns and that, whenever the number is written and whatever its size, it will always be bounded either by blanks or by the beginning or end of the model output file line. If only blanks occur between (and including) the two listed columns, or if non-numeric characters or two number fragments are found, an error condition will occur and the UCODE_2005 will terminate with an appropriate error message.

As for fixed dependents, it is normally not necessary to have secondary markers, whitespace, and tabs on the same line as a semi-fixed dependent, because the column numbers provided with the semi-fixed dependent instruction determine the location of the dependent on the line. If more than one semi-fixed dependent instruction is provided on a single UCODE_2005 instruction line, the column numbers pertaining to these dependents need to increase from left to right.

It takes more computer effort to read a semi-fixed dependent than a fixed dependent, so fixed dependents are preferable.

After a semi-fixed dependent is read, the cursor is positioned at the end of the number extracted. Any further processing of the line must take place to the right of that column.

Non-Fixed Reading

Figure 13 demonstrates the use of non-fixed reading.

When a non-fixed reading instruction is encountered, UCODE_2005 searches forward from its current column until it finds a non-space; it assumes this character is the first character read. Then it again searches to the right until it finds either a space, the end of the line, or the first character of a secondary marker which follows the non-fixed reading instruction in the instruction file. In this way it identifies a string of characters and tries to read them as a number; if this is unsuccessful due to the presence of non-numeric characters or some other problem, execution terminates with an appropriate error message. An error condition will also arise if the end of a line is encountered while looking for the beginning of a non-fixed reading.

Consider the portion of the model output file shown in the top part of Figure 13. The species populations at different times cannot be read as either fixed or semi-fixed dependents, because the numbers representing these populations do not consistently fall within one range of column numbers on the model output file. Figure 13 shows how non-fixed reading can be used in this situation.

Portion of model output file	. SPECIES POPULATION, 1 YEAR = 1.23498E5 SPECIES POPULATION, 2 YEARS = 1.58374E5 SPECIES POPULATION, 3 YEARS (ITERATION NEEDED)= 1.78434E5 SPECIES POPULATION, 4 YEARS = 2.34563E5 .
Instruction file	┤if * . *SPECIES POPULATION, * * = * !sp1! l1 * = * !sp2! l1 * = * !sp3! l1 * = * !sp4! .

Figure 13. Portion of the model output file and instruction file illustrating use of primary and secondary markers and non-fixed reading. Numbers being read are shaded. Dots replace one or more lines that are not shown.

A primary marker is used to move the cursor to the first of the lines shown in Figure 13. Then, the " = " string is used as a secondary marker. After it processes a secondary marker, the cursor resides on the last character of that marker, in this case on the blank following the "=". Hence after reading the " = " string, the !sp1! instruction is processed by isolating the string "1.23498E5".

After reading the model-calculated value for "sp1", the next instruction line is executed. In accordance with the "l1" (line 1) instruction, the next line of the model output file is read into memory, then searched for an "=" character and the next number is read as "sp2". The procedure is then repeated to read "sp3" and "sp4".

Successful identification of a non-fixed dependent depends on the instructions preceding it. The secondary marker, tab, and whitespace instructions will be most useful in this regard, though fixed and semi-fixed reading may also precede a non-fixed reading; in all these cases the cursor is placed over the last character of the string or number it identifies on the model output file corresponding to an instruction item, before proceeding to the next instruction.

Figure 14 further illustrates the use of non-fixed reading. To read the fourth number when the lengths of the preceding numbers are unknown, a non-fixed reading is needed.

Here it is assumed that, prior to reading this instruction, the cursor was located on the 10th preceding line of the model output file. As long as it is known that no space will ever precede the first number, there will always be three incidences of whitespace preceding the number that must be read. However, if it happens that one or more spaces may sometimes precede the first number, then the first number can be read as a dummy reading as shown in instruction file 2 in Figure 14.

A third alternative for locating the value to be read for "obs1" in Figure 14 is to use the dummy read more than once. Had the numbers been separated by commas instead of spaces, the commas would need to be used as secondary markers to find "obs1".

Portion of model output file	. . 4.33 -20.3 23.392093 `3.394382` . .
Instruction file 1	l10 w w w !obs1!
Instruction file 2	l10 !dum! w w w !obs1!
Instruction file 3	l10 !dum! !dum! !dum! !obs1!

Figure 14. Portion of a model output file and instruction file illustrating use of white space for non-fixed reading. The number being read is shaded. Dots replace one or more lines that are not shown. Instruction file 1 works if the line never begins with a space. Instruction files 2 and 3 work whether or not the line begins with a space.

A number not bounded by blanks or the beginning or end of a line can still be read as a non-fixed dependent with the proper choice of secondary markers. Consider the model output file line shown in Figure 15

Portion of model output file	. . SOIL WATER CONTENT (NO CORRECTION)=`21.345634`% .
Instruction file	jif * . . l5 *=* !sws! *%*

Figure 15. Portion of a model output file and instruction file illustrating use of a secondary marker to define the end of a non-fixed reading. The number being read is shaded. Dots replace one or more lines that are not shown.

It is not possible to read the soil water content as a fixed dependent because the "(NO CORRECTION)" string may or may not be present after any particular model run. Without the secondary marker the extracted string would include the "%" character and a run-time error would occur. Secondary markers that are used to terminate the reading of a number can not begin with a number.

The lack of a space between the "=" character and the number that must be read is not problematic. After processing of the "=" character as a secondary marker, the cursor falls on the "=" character itself. The search for the first non-space initiated by the !sws! instruction terminates on the very next character after the "=", that is, the "2" character. Then this character is accepted as the left boundary of the extracted number and search proceeds forward for the right boundary of the string in the usual manner.

After a non-fixed reading, the cursor is placed on the last extracted character. It can then undertake further processing of the model output file line to read further non-fixed, fixed, or semi-fixed dependents, or process navigational instructions as directed.

Making an Instruction File

An instruction file is easily built using a text editor. Normally two windows should be open – one to write the instructions, and the other to view the model output file.

Performance of an instruction file needs to be checked thoroughly before proceeding.

Some errors are obvious. For example, if no number is found or a letter is read instead of a number, a run-time error occurs and an error message is written that identifies the problematic command.

Some errors are not obvious. For example, if a number is read but it is wrong. Indeed, it may only be apparent if the UCODE_2005 appears to be malfunctioning, for example if it reports an unusually high objective function that cannot be lowered. The problem may be difficult to identify through consideration of the simulated equivalents printed in tables that compare simulated equivalents to observed values, because those simulated values may be a function of one or more extracted values. To help find these kind of errors set verbose=3 or 4 in the UCODE_CONTROL_DATA input block so that the extracted values are printed in the main output file.

Chapter 12: INPUT FOR PARALLEL EXECUTION

UCODE_2005 can take advantage of parallel execution at two levels: (1) the process model(s) can be programmed and set up to use multiple processors and (2) UCODE_2005 can be set up to send runs of the command line defined in the Model_Command_Lines input block to different processors.

The first type of parallelization generally is advantageous if sensitivities are produced within a run of the process model(s), and is discussed by process models that support the use of multiple processors, such as MODFLOW-2000.

The second type of parallelization is advantageous if sensitivities are calculated by UCODE_2005 using perturbation methods. In this chapter we use this situation to explain how parallel processing can be used to reduce execution time.

In some circumstances some of the sensitivities may be calculated by the process model while others are calculated by UCODE_2005 using perturbation methods. Such calculation can be parallelized easily by sending the process model runs that calculate sensitivities to one computer while the other runs of the process model are sent to other computers.

Using Multiple Processors to Calculate Perturbation Sensitivities

The computation time attained using multiple processors depends on the number of parameters for which sensitivities are calculated (equivalent to the number of parameters being estimated if performing parameter estimation), the type of sensitivities calculated, and the number of processors used. To help users approximate execution times, here we consider the situation in which perturbation sensitivities are calculated using the forward-difference method. In this situation, the number of process-model solutions required for each parameter-estimation iteration equals the number of parameters plus one for the base run. On one processor, execution time is approximately (Hill and Tiedeman, in press, Chapter 15.1; Hill, 1998, 66):

$$[\text{the time required for a forward solution}] \ [1 + \text{the number of parameters}]. \tag{9}$$

If enough processors are available, execution time can be reduced to:

$$2 \times [\text{the time required for a forward solution}]. \tag{10}$$

The factor of 2 results because the base run is run first and by itself. Though this is not required for forward-perturbation sensitivities, it is how many programs, including UCODE_2005, perform the calculations. For twenty parameters, execution times would be reduced by about a factor of 10 if 20 or more processors of about equal speed were available.

If fewer processors are available and one processor is used only for the base run and to coordinate other processors, execution times are approximately proportional to

[the time required for a forward solution] $\times$

(11)

[1 + (number of parameters)/(number of processors - 1)],

Any fraction in the division results in the number being equal to the next larger integer. If all processors have the same execution speed, they are used most efficiently if the number of processors minus one divides evenly into the number of parameters. For example, if four processors are used and nine parameters are defined, the number in parentheses will be three and all processors will be used nearly continuously. For the same nine parameters and five processors, the number in parentheses will still be three so that the expected time of solution will be the same; three of the processors will be idle while one performs calculations for the eighth parameter. The situation is shown in Table 10.

Table 10: The sequence of calculations performed by UCODE_2005 given nine parameters and (A) four or (B) five computer processors of about equal speed.

(A)

Processor Number	The process model is run with the base set of parameter values and then with the value of the indicated parameter perturbed using four processors.		
1	Base run. Coordinate the other processors		
2	Parameter 1	Parameter 4	Parameter 7
3	Parameter 2	Parameter 5	Parameter 8
4	Parameter 3	Parameter 6	Parameter 9

(B)

Processor Number	The process model is run with the base set of parameter values and then with the value of the indicated parameter perturbed using five processors.		
1	Base run. Coordinates the other processors		
2	Parameter 1	Parameter 5	Parameter 9
3	Parameter 2	Parameter 6	(idle)
4	Parameter 3	Parameter 7	(idle)
5	Parameter 4	Parameter 8	(idle)

If the processor speeds differ or some of the runs take more execution time than others, UCODE_2005 adjusts the runs assigned to the different processors.

If a run fails on any processor, a message to that effect is printed to the main UCODE_2005 output file and the processor is taken off the list of processors to use. The run then gets assigned to the next available processor. If the failure is caused by model input (for example, unrealistic parameter values), successive processors will fail as a result of trying to make the run with the problematic model input. Eventually, all processors will fail, and execution of UCODE_2005 will stop.

Execution time can be improved by also using processor 1 for model runs. This works best when, on processor 1, the priority of UCODE_2005 is set to something less than the default priority. On windows operating systems, this can be accomplished as follows. Start UCODE_2005. Press cntl-alt-delete to get a menu and then click on "Task Manager." Click the tab "Processes". Click on the column header "Image Name" twice so that UCODE_2005 is listed toward the top. Along the top of the window, click on "View" to get a drop-down menu, and click "Select Columns". In the right column, make sure the box to the left of "Base Priority" has a check in it. Click "OK" to get back to the list of processes. Enlarge the window as needed to view the column labeled "Base Priority" Locate UCODE_2005 and right click on it. Click "Set Priority" and choose "BelowNormal".

Parallel Processing Using the Dispatcher-Runner Protocol

The dispatcher-runner protocol used here is a version of what is sometimes called the master-slave protocol. The dispatcher-runner protocol can be used to implement parallel processing using multiple processors. In practice the multiple processors often are in multiple networked computers, but can also be on one multi-processor computer.

To set up parallel processing, identify a directory to use on each processor (if the dispatcher and a runner are to be on one processor, two directories are needed there). Select the directory from which to run the dispatcher program (here, UCODE_2005). Directories to be used for process model runs are called runner directories. UCODE_2005 needs to be able to write and read files in each runner directory. For each runner directory, copy in all the files and directories needed to run the process model(s) and the program JRUNNER. Start at least one JRUNNER before stating UCODE_2005. Additional instances of JRUNNER may be started after UCODE_2005 is started.

Communication between UCODE_2005 and the runner programs is implemented by creating, writing, reading, and deleting signal files. Table 11 lists the signal files and how they are used. A conflict exists if a runner directory contains a file with the same name as a signal file, so the file names listed in Table 11 can not be used for other purposes.

The performance of the dispatcher-runner parallel processing protocol is described in the following paragraphs. In this description UCODE_2005 is the dispatcher program.

When JRUNNER starts, it writes signal file "jrunner.rdy" to indicate its active status to the dispatcher program. It then begins looking in its directory for signal files written by the dispatcher program. The dispatcher program executes the overall logic of the program, as in the case without the parallel-processing capability. The dispatcher program writes a signal file named "jdispatcher.rdy" in each runner directory to communicate initialization information, which is then read by JRUNNER.

At a point in the dispatcher program where a set of model runs is required that can be executed in parallel, a signal file named "jdispar.rdy" is written to the directory of each active runner. Each copy of "jdispar.rdy" contains a set of parameter values and other data needed for a single model run. When JRUNNER detects the existence of a jdispar.rdy file in its directory, it reads the parameter values and other data from the file. JRUNNER uses the parameter values and template files to prepare one or more model-input files and then sends a command to the operating system to initiate a model run. When the model run is complete, JRUNNER extracts model-calculated values from one or more model-output files and writes the values to a signal file named "jrundep.rdy".

The dispatcher program iteratively checks the directory of each of the active runners for the presence of a file named "jrundep.rdy"; when it finds this file it reads the model-calculated values from it.

Table 11. Signal files for parallel processing

File name	Program that writes file	Purpose
jrunner.rdy	JRUNNER	Inform dispatcher that runner is active and ready to make a model run.
jdispatcher.rdy	Dispatcher	Communicate to runner data needed to start model runs.
jdispar.rdy	Dispatcher	Communicate to runner: command to be used to start a model run, parameter values to be used, and that runner should initiate a model run.
jrundep.rdy	JRUNNER	Communicate model-calculated dependent values from runner to dispatcher.
jdispatcher.fin	Dispatcher	Stop execution of runner.
jrunner.fin	JRUNNER	Communicate to dispatcher that signal to stop execution has been received, and that runner is stopping.
jrunfail.fin	JRUNNER	Communicate to dispatcher that runner has encountered an error and is stopping.

All instances of JRUNNER prepare model-input files and extract model-calculated values from model-output files in the same way. For this approach to work correctly, the following requirements need to be met:

1. Avoid file-access conflicts. Set up the runner directories and all directories containing files used by JRUNNER and the model to be independent of each other.

2. The runner directories need to be accessible from the directory where the dispatcher program runs using pathnames, on a computer or within a network. The pathnames can be absolute or relative. In computing environments that support it, the Universal Naming Convention (UNC) may be used to construct pathnames.

3. For assured results, run the dispatcher program and all instances of JRUNNER under a common operating system. It may be possible to obtain successful results using a combination of closely related operating systems, such as different versions of UNIX, or a combination of UNIX and Linux. Experimentation is needed to determine if a particular combination of operating systems can be successfully used in parallel processing. The critical issue that determines if operating systems can be used in combination is that text files written by the dispatcher program need to be readable by all of the runner programs, and vice versa. This restriction, for example, likely would prevent a dispatcher program to run on a Windows computer with

JRUNNERs running on UNIX computers, because the line-ending conventions used by the two systems are different.

When JRUNNER encounters an error that it is designed to handle, it communicates an error message to the dispatcher program in the "jrunfail.fin" file and stops execution. When this situation arises, UCODE_2005 writes the error message to the screen.

The methodology used by the parallel-processing capability implemented in the current version of the Parallel-Processing Module has both positive and negative attributes. It is straightforward to set up parallel processing because no additional software is needed. The method is fairly robust in that failure of individual runners can be tolerated, failed runners can be restarted without restarting the dispatcher program, and new runners can be added after the dispatcher program has started.

A major drawback of the method is that both the dispatcher program and JRUNNER use the central processing unit (CPU) of the computer virtually constantly while they are running, because they continually are either running the model or testing for the existence of signal files. This attribute makes the use of the parallel-processing capability unwelcome in a computing environment shared by multiple users. However, given exclusive use of multiple computers, the savings in execution times can be substantial.

In the simplest case, instances of JRUNNER are started manually, by typing "JRUNNER" at the command prompt or by invoking JRUNNER from a script or batch file. Alternatively, utilities for starting a program remotely could be used to make the starting of JRUNNER instances more convenient. However, the issue of starting programs remotely is outside the scope of UCODE_2005 and the JUPITER API.

In some circumstances better performance can be achieved by reducing the priority of the JRUNNER relative to the process model. On Windows operating systems, this can be accomplished using the method described at the end of the previous section because JRUNNER runs continuously after it is started until the end of the UCODE_2005 run.

Parallel_Control Input Block (Optional)

The Parallel_Control input block defines how parallel processing is accomplished.

The keywords are:

Parallel
- yes: Activate parallel processing. no: one computer is used for the computations. Default=no.

Wait
- Time delay, in seconds, used in file management. Default=0.001. Increase if the message "Warning: WAIT time may be too small" is printed to the screen of any of the computers being used.

Opening, writing, reading, and closing of disk files are processes that require a finite amount of time. The current parallel-processing capability of the Parallel-Processing Module communicates through disk files, and a certain amount of delay time is built into the sections of code that deal with these files. The WAIT variable of the Parallel_Runners input block provides the user with a way to control the increment of time used in these sections of code. Generally, a small fraction of a second is sufficient—the default value of WAIT is 0.001 second. WAIT needs to be set to a larger value if the message "Warning: WAIT time may be too small" appears on the screen running either the dispatcher program or JRUNNER. Little is gained by setting WAIT smaller than the default value.

VerboseRunner
- Flag that controls printing by the runner. Problems with parallel processing often can be diagnosed using messages written by JRUNNER. Default=3.

VerboseRunner	Output
0	No extraneous output.
1	Warnings.
2	Warnings, notes.
3	Warnings, notes, echo selected input.
4	All optional messages are printed.

AutoStopRunners
- yes: execution of runners will be stopped when execution of UCODE_2005 stops. no: when execution of UCODE_2005 stops, runners will reset themselves in preparation for another execution of UCODE_2005. Default=yes.

OperatingSystem
- String identifying the operating system. All computers need to use the same operating system. OperatingSystem may be 'Windows', 'DOS', 'Unix', or 'Linux'. This variable affects the operating-system command used to rename files in the runner directories. Default=Windows.

TimeoutFactor
- Factor that multiplies RUNTIME to determine if a model run is overdue. Default=3.0.

Chapter 12: Input to Execute the Process Model(s) in Parallel
            --Parallel_Control Input Block--

If Blockformat TABLE format is selected, COLUMNLABELS are needed because there
is no default column order for the Parallel_Control input block.

Example Parallel Control input block:

```
BEGIN PARALLEL_CONTROL
 OPERATINGSYSTEM=Linux
 Parallel=yes
 Wait=.001
 VerboseRunner=3
 AutoStopRunners=yes
END PARALLEL CONTROL
```

Parallel_Runners Input Block (Optional)

The Parallel_Runners input block defines how parallel processing is accomplished. This input block is expected to include the following items for each runner.

The keywords are:

RunnerName — Name by which runner is identified. If the name includes spaces it needs to be enclosed in single quotes. The name can be up to 20 characters long.

RunnerDir — Pathname of directory where the runner program runs. The path needs to end with a backward or forward slash (\ or /), depending on the convention used by the operating system.

RunTime — Expected model runtime, in seconds. Default=10.

If JRUNNER encounters an unrecognized error, or stops due to an external problem such as a power interruption or network failure, no signal file is generated by JRUNNER to communicate the problem to the dispatcher program. The Parallel-Processing Module handles situations like this by comparing the elapsed time since writing the jdispar.rdy file to an expected run time. If the elapsed time exceeds the expected run time by a factor of TimeOutFactor (from the Parallel_Control input block), the model run is assumed to have failed, that instance of JRUNNER is assumed to be inactive, and the Dispatcher stops checking for results. The model run that failed is then assigned to the next available, active runner. The RunTime variable of the Parallel_Runners input block allows the user to specify an initial expected run time. As model runs are completed by each active runner, the expected run time is adjusted, so that the expected run time will approach the actual run time on each runner. Because the expected run times are continually being adjusted and a run is not assumed to have failed until the elapsed time exceeds 3 times the expected run time, RunTime does not need to be an accurate estimate of the model run time.

If Blockformat KEYWORDS is selected by designation or default, keywords defining a Runner in the Parallel_Runners input block need to be grouped together and follow the related RunnerName. The RunnerName keyword needs to be the first keyword on a new line. RunnerName and associated keywords are repeated to define multiple runners.

If blockformat TABLE is selected without indicating ColumnLabels, the default column order is used. No columns are ignored and a column for each keyword is needed. If ColumnLabels are indicated, the column labels can appear in any order; the RunnerName keyword need not be first, though it often is first.

Default Column Order:
```
RUNNERNAME RUNNERDIR RUNTIME
```

Chapter 12: Input to Execute the Process Model(s) in Parallel
        --Parallel_Runners Input Block--

Example Parallel Runners input block:

```
BEGIN PARALLEL_RUNNERS TABLE
# RUNNERDIR must end with the correct directory separator for
#   the OS -- "\" for Windows and "/" for Unix and Linux.
NROW=6  NCOL=4  COLUMNLABELS
RUNNERNAME RUNNERDIR      RUNTIME
runner1    \dir\runner1\   8000
runner2    \dir\runner2\   8000
runner3    \dir\runner3\   8000
runner4    \dir\runner4\   8000
runner5    \dir\runner5\   8000
runner6    \dir\runner6\   8000
END PARALLEL RUNNERS
```

Chapter 12: Input to Execute the Process Model(s) in Parallel
--Parallel_Runners Input Block--

Chapter 13: EQUATION PROTOCOLS AND TWO ADDITIONAL INPUT FILES

Equation Protocols for the UCODE_2005 Main Input File

In UCODE_2005, equations can be defined in the following input blocks:
For parameters: Derived_Parameters,
For observations: Observation_Data, Derived_Observations,
For predictions: Prediction_Data, and Derived_Predictions.

Equations are given a name and a mathematical expression which represents the right side of the "=" sign. The equal sign is not included in the expression. The expression can consist of the arithmetic operator and functions listed in Table 12 and variable names (for example, parameter names or observation names). Equations that include spaces need to be surrounded by single or double quotes.

The order in which mathematical operations are carried out to evaluate a mathematical expression is the same as that used in normal mathematical operations. That is, raise to a power, followed by multiplication and division, followed by unary addition and subtraction, followed by binary addition and subtraction. Parentheses can be used to override or clarify this order.

Example Equations

The following are some examples of acceptable equations. It is assumed that values are available for variables parname1, parname2, obsname1, and obsname2.

```
"parname1 + sqrt(parname2*parname1)"

sqrt(abs(sin(obsname1/57.29)))

"exp(3.0 * sqrt(obsname1/obsname2))"

parname1

1.0
```

These examples demonstrate the following:

1. Spaces can be left between operators, variable names, brackets, and so on if the equation is enclosed in double or single quotes. However, a variable name can not include a space.

2. An equation entity that is not an operator or a function is first treated as a number. If it cannot be read as a number, it is assumed to be a variable. To avoid confusion, variable names can not begin with a number.

3. If an illegal argument is supplied to any function (for example if a negative number is provided as the argument to a log or sqrt function), an error condition arises, the error is reported, and UCODE_2005 execution stops.

Table 12. Arithmetic operators and functions available for equations.

Arithmetic Operator	Operation
** or ^	Power. $a**b$ or a^b is interpreted as "a raised to the power b".
/	Division. a/b is interpreted as "a divided by b".
*	Multiplication. $a*b$ is interpreted as "a multiplied by b".
-	Subtraction. This can be a unary or binary operator. $a-b$ is interpreted as "a minus b"; $-a$ is interpreted as "negative a".
+	Addition. This can be a unary or binary operator. $a+b$ is interpreted as "a plus b"; $+a$ or a is interpreted as "positive a".
( )	Parentheses. Terms in parentheses are evaluated first. For example: $5 + 4 * 3$ is evaluated as 17. However $(5 + 4) * 3$ is evaluated as 27.
Function	**Definition**
abs( )	Absolute value. Argument can be any floating-point number.
cos( )	Cosine. Argument can be any floating-point number supplied in radians.
acos( )	Inverse cosine. Absolute value of argument must be less than or equal to one. Value is returned in radians.
sin( )	Sine. Argument can be any floating-point number supplied in radians.
asin( )	Inverse sine. Absolute value of argument must be less than or equal to one. Value is returned in radians.
tan( )	Tan. Argument can be any floating-point number supplied in radians.
atan( )	Inverse tan. Argument can be any floating-point number. Value is returned in radians.
cosh( )	Hyperbolic cosine. Argument can be any floating-point number.
sinh( )	Hyperbolic sine. Argument can be any floating-point number.
tanh( )	Hyperbolic tan. Argument can be any floating-point number.
exp( )	Exponential. Argument can be any floating-point number.
log( )	Log to base e. Argument must be a positive floating-point number.
log10( )	Log to base 10. Argument must be a positive floating-point number.
sqrt( )	Square root. Argument must be non-negative.
min(, ,)	Minimum of a series of numbers. Arguments can be any floating-point numbers.
max(, ,)	Maximum of a series of numbers. Arguments can be any floating-point numbers.
mod(,)	Remainder. $mod(a,b)$ is the remainder after a is divided by b.

Derivatives Interface Input File

A Derivatives Interface input file provides UCODE_2005 with information needed to obtain model-calculated sensitivities (derivatives of simulated values with respect to parameters) from a model-output file rather than calculating them by perturbation.

The Derivatives Interface file does not provide as much flexibility as calculating sensitivities by perturbation. The following restrictions apply:

1. Sensitivities can only be read for observations and predictions with UseFlag=yes. In some circumstances, lines or columns of sensitivities for other simulated values can be omitted using DERFORMAT of the Derivatives Interface input file.

2. Sensitivities can only be read for adjustable parameters – that is, parameters with Adjustable=yes in the Parameter_Groups or Parameter_Data input block. In some circumstances, lines or columns of sensitivities for other parameters can be omitted using DerFormat of the Derivatives Interface input file.

3. No equations can be applied to derivatives read using the Derivatives Interface file. This includes equations related to parameters, observations, and predictions. This is because equations are applied by UCODE_2005 to simulated values, not sensitivities. Sensitivities for parameters, observations, and predictions that are calculated using equations and referred to as being derived in this work need to be calculated using perturbation.

4. For a single parameter, sensitivities for all observations or predictions need to either be read using a Derivatives Interface file or to be calculated by perturbation by UCODE_2005. Calculating sensitivities for the same parameter using different methods is not supported because it increases execution time and is rarely advantageous.

To use a Derivatives Interface input file, the following are needed in the main input file for UCODE_2005:

1. In the Options input block, name a Derivatives Interface input file using the Derivatives_Interface keyword.

2. In the Model_Command_Lines input block, the keyword Purpose needs to be Derivatives or Forward&Der for at least one occurrence of the Command keyword.

3. In the Parameter_Data input block, SenMethod needs to equal -1 or 0 for parameters for which sensitivities are read using the Derivatives Interface input block. SenMethod=-1 indicates that sensitivities for log-transformed parameters are read as transformed sensitivities. SenMethod=0 indicates that sensitivities for

log-transformed parameters are read as native parameter sensitivities and are transformed by UCODE_2005.

In addition, the Derivatives Interface input file needs to be constructed as described in Table 13.

Table 13. Derivatives Interface file input instructions.
[The items need to be in the order shown. All items are read in free format. Sensitivities can be for native parameter values or log-transformed as in the regression; see point 3 preceding table 13.]

Item	Variable Name (bold) or content	Explanation
0	# Text	Zero or more comment lines allowed only at the top of the file. Comments are identified by # in column 1. Comments may be inserted on data lines following the required data, except not on the lines containing names of parameters or dependents.
1	**DERFILE**	Name or path of the model-generated file containing derivatives
2	**NSKIP**	Number of lines to skip at the top of DERFILE before reading derivatives
3	**NDEP NPAR**	Number of simulated values and parameters for which sensitivities are to be read from DERFILE.
4	**ORIENTATION**	Either "**ROW/DEP**" or "**ROW/PAR**", with or without quotes.[1]
5	**DERFORMAT**	Fortran format for reading derivative values, or "**(FREE)**".[2]
6	**"PARAMETERS"**	Enter the word "PARAMETERS", with or without quotes. Interpretation is not case-sensitive.
7	Parameter names	NPAR parameter names. The names need to correspond to and be in the same order as the parameters for which model-calculated derivatives are provided in file DERFILE.[3]
8	**"DEPENDENTS"**[4]	Enter the word "DEPENDENTS"[4], with or without quotes. Interpretation is not case-sensitive.
9	Any combination of Obsname Predname	NDEP names of simulated values for which sensitivities are listed. The names need to correspond to and be in the same order as the dependents for which model-calculated derivatives are provided in file DERFILE.[3]

[1] ORIENTATION: "ROW/DEP" indicates that each row in file DERFILE contains derivatives for one simulated (dependent) value. "ROW/PAR" indicates that each row contains derivatives for one parameter.

[2] DERFORMAT: The format string needs to include the parentheses. Single or double quotes may be used to include embedded spaces or commas. Length limit: 200 characters. The format is executed once for each row of data.

[3] Parameter names and Obsname or Predname: Names are read until NPAR (or NDEP) names are read, the names should be in the order used in the model-generated derivatives file. Multiple names may be listed on each line. The names need to correspond to parameter or dependent names defined elsewhere in program input.

[4] The term 'dependent' refers to values simulated by the model. They may be identified as
    observations or predictions elsewhere in UCODE_2005, but the distinction is not needed here
    when reading the derivatives interface file.

Example of a derivatives interface file:

```
# Derivatives Interface File for u test model
# Parameter and dependent names can be listed in a row or as a
# column of labels
ex. su           (DERFILE)
1                (NSKIP)
11   5           (NDEP NPAR)
row/dep          (ORIENTATION)
'(21x,5f15.0)'   (DERFORMAT)
PARAMETERS
k1              KC              k2m             RCH1            RCH2
DEPENDENTS
A1 B1 C1 D1 E1 A2 B2 C2 D2 E2 flow
```

fn.xyzt Input File

The fn.xyzt input file can be used in the forward, sensitivity analysis, and parameter – estimation modes to facilitate plotting of weighted residuals and residuals. The plots may be spatial and(or) temporal and may be maps, cross-sections, hydrographs, and so on. The filename begins with the filename prefix defined on the command line, which is referred to as fn in this document. UCODE_2005 searches for a file with this name and if it is found UCODE_2005 produces a data-exchange file with file extension _xyztwr.

The first line of the xyzt input file is ignored and can be used to list column headings, projection information, or other information. The rest of the file needs to be composed of lines containing five columns of data:

 OBSERVATION NAME   X   Y   Z   TIME

As long as at least one space or a comma follows the time, additional data or comments to the right are ignored.

Typically all possible observations are listed in the fn.xyzt input file. For any single execution of UCODE_2005, all observations with UseFlag=yes need to be included in the file with the ObsName correctly typed for the _xyztwr file to be produced. If there are observations with UseFlag=yes that are not listed in the fn.xyzt file the following occurs: (1) an error message is written to both the screen and the output file and (2) the _xyztwr data-exchange file is not printed. Execution of UCODE_2005 continues.

Observations listed in the fn.xyzt file are ignored if they are not used in the run of UCODE_2005. This is intended to allow users to investigate regressions with different sets of observations without having to modify the fn.xyzt file.

The fn.xyzt input file is used only to create the fn._xyztwr file; the data are not used in the UCODE_2005 calculations. The fn.xyzt file is intended to provide plotting positions to assist in evaluating model fit. For features that do not occur at a point, such as long well screens and stream reaches, enter the location at which values associated with the observation or prediction are to be printed.

UCODE_2005 produces the _xyztwr file because plotting of weighted and unweighted residuals in relation to space and time is so fundamental to model calibration. The fn.xyzt file can be used with other types of information such as sensitivities to produce graphics. For predictions, a file similar to the xyzt file may be constructed for use with the data-exchange file that lists simulated predictions. These other types of graphs can easily be produced by reading the xyzt file and other file of interest into a plotting routine.

Chapter 14: UCODE_2005 OUTPUT FILES

All UCODE_2005 output files are named using the filename prefix defined on the command line and referred to in this documentation as fn.

If DataExchange=yes in the UCODE_Control_Data input block, then extensive information is printed to data-exchange files. Data-exchange file names are of the form: fn._xxx, where _xxx is called a filename extension and xxx is replaced by one or more characters. For UCODE_2005, the options for _xxx and the contents of the resulting files are described in Table 15 through Table 21.

The primary output file for UCODE_2005 is named fn.#xxx, where #xxx is called the filename extension and xxx is replaced by a label that depends on the mode, as mentioned in Table 3 and in table 20. There are a variety of opportunities to choose more or less information to be printed to this file, as listed in Table 14.

Output files and data-exchange files with similar naming conventions also are produced by the post-processing programs documented in this report. To help the user readily identify these files, a complete list of files is presented alphabetically by file extension in Appendix B.

An example of the UCODE_2005 main output file for parameter-estimation mode and selected data-exchange files are shown in Appendix C.

Table 14. Input variables available to control UCODE_2005 output.
[Except for DataExchange, the variables affect the UCODE_2005 main output file.]

Input Block	Keyword	Output controlled
Options	Verbose	Warnings, echoed input, and other. Set to 4 or 5 to check the numbers read using the instructions of the Model_Output_Files input block.
UCODE_Control_Data	DataExchange	Data-exchange files
	StartRes IntermedRes FinalRes	Controls printing of observed and simulated values and residuals calculated using starting, intermediate[1], and final parameter values.
	StartSens IntermedSens FinalSens	Controls the types of sensitivity tables printed using starting, intermediate[1], and final parameter values.

[1] Intermediate parameter values are calculated for each parameter-estimation iteration.

Main UCODE_2005 Output File

The name of the UCODE_2005 main output file always begins with the filename prefix defined on the command line and referred to in this document as fn. The file extension used for the main output file depends on the mode (Table 3); it always begins with the character "#". For modes Forward, Sensitivity Analysis, and Parameter Estimation, the main UCODE_2005 output file is fn.#uout. For prediction mode, the main output file is fn.#upred. For test-model-linearity mode, the main output file is fn.#umodlin. The main output files from other modes are discussed in Chapter 17.

The main output file contains an echo of the input data, followed by selected results depending in the mode (Table 3) and the chosen options (Table 14).

Data-Exchange Files Produced by UCODE_2005

Data-exchange files are printed when DataExchange=yes in the UCODE_Control_Data input block. Data-exchange files produced by UCODE_2005 are listed in Table 15 and B-1. Their connection to the JUPITER API is discussed in Appendix A.

As mentioned in the introduction of this report, the filenames of data-exchange files are created using the filename prefix defined on the command line and an extension that begins with an underscore. Often the underscore is followed by two letters, with one letter indicating a quantity in the file that commonly would be plotted on the y-axis followed by one letter indicating a quantity that commonly would be plotted on the x-axis. This order is consistent with how the contents of a graph are often described. In such cases, the columns in the file are ordered with the x-axis first because that is the order expected by most plotting programs. For example, the data-exchange file with extension _os is generally used to create a graph of observed and simulated values and contains a column of simulated values followed by a column of observations and prior information. Alternatively, the names are descriptive. For example, the file with extension _sd contains dimensionless scaled sensitivities. An s follows the underscore for all files containing sensitivities.

Also as mentioned in the introduction, data-exchange files are designed to facilitate the exchange of data among computer codes and post-processing software. For most of these files the only explanatory text are labels on the first line that identify subsequent columns of data. Each label is enclosed in double quotes. Two files contain more than one table of data; these have filename extensions _pa and _pasub. Two files contain matrices listed in the compressed format described in Chapter 10; these have filename extensions _wt and _wtpri.

The contents of the data-exchange files produced by UCODE_2005 are described in tables 16 through 21.

Table 15. Brief description of data-exchange files produced by UCODE_2005.
[Modes that produce the files are listed. SEN, Sensitivity Analysis; PE, Parameter Estimation.]

File Extension	Content (also see tables 16-21)
Files used for model development and testing	
_dm	Information related to model structure, fit and parsimony.
_gm	Defines observation groups.
_pr	Prior information equations.
Information from each parameter-estimation iteration [Mode: Sen[1], PE]	
_pa	All defined parameters: values formatted in columns.
_pasub	All defined parameters: formatted to substitute in Parameter_Values input block.
_ss	Sum of squared weighted residuals.
Final results [Mode: Sen[1], PE]	
_paopt	All defined parameters: See table 19 for content.
_pc	Adjusted parameters: See table 19 for content
Graphical analysis of model fit to observations and prior information [Mode: Forward, Sen, PE]	
_nm	Weighted residuals and probability plotting positions.
_os	Unweighted simulated equivalents and observed or prior values.
_r	Unweighted residuals.
_sos	Parameter values and resulting value of sum-of-squares objective function.
_w	Weighted residuals for observations and prior information.
_ws	Simulated equivalents and weighted residuals.
_ww	Weighted simulated equivalents and weighted observations or prior information.
xyztwr	Merger of _r and _w data-exchange files with the optional xyzt input file.
Commonly used for sensitivity analysis [Mode: Sen, PE]	
_pcc	Large parameter correlation coefficients ($\geq$0.85) formatted for presentation.[2]
_sc	Composite scaled sensitivities for each parameter.
_sd	Dimensionless scaled sensitivities for each observation and each parameter.
_so	Leverage statistics with observations and, if defined, prior information.
_s1	One-percent scaled sensitivities. (applicable in limited circumstances)
Used by other programs; rarely by users [Mode: Sen, PE, and as noted]	
_b1	Parameter sets for calculating Beale's measure of linearity.[3]
_b2	Simulated equivalents for Beale parameter sets of _b1.[3]
_init	Like _paopt but for a non-optimal parameter set.[4]
init.**	** is replaced by dm, mv, su, or supri.[4]
_mc	Parameter correlation coefficient matrix.[2]
_mv	Parameter variance-covariance matrix.[5]
_su	Unscaled sensitivities for observations.
_supri	Unscaled sensitivities for prior information.
_wt	Weighting for observations. [Mode: also Forward]
_wtpri	Weighting for prior information. [Mode: also Forward]
Files for prediction analysis [Mode: Prediction]	
_dmp	Number of predictions.
_gmp	Defines prediction groups.
_p	Predicted values.[5]
_pv	Prediction variances.[5,6]
_spu	Unscaled sensitivities for predictions.[5]

Table 15.-- Continued

File Extension	Content (also see tables 16-21)
Prediction scaled sensitivities. For each, one value is printed for each prediction and parameter.	
_sppp	Scaled using parameter values and predicted values.
_sppr	Scaled using parameter values and reference values. [7]
_spsp	Scaled using parameter standard deviations and predicted values.
_spsr	Scaled using parameter standard deviations and reference values. [7]

[1] Includes the results for the parameter values listed using keyword StartValue in the UCODE_2005 main input file.

[2] File contents are meaningful for (a) sensitivity-analysis mode or (b) parameter-estimation mode if convergence is reached or the maximum number of iterations is reached and the Reg_GN_Controls input block Stats_On_Nonconverge=yes by designation or default.

[3] Used by post-processing program MODEL_LINEARITY. The _b2 file is produced by test-model-linearity mode.

[4] The _init._** file is like the _** file but contents are calculated using a non-optimal parameter set. The _init and _init._** data-exchange files are not listed in tables 16 to 21. They have the same header lines as the associated files listed here. These files are used by the UCODE_2005 nonlinear-uncertainty mode. See Chapter 17.

[5] Used by post-processing program LINEAR_UNCERTAINTY.

[6] Calculated using MeasStatistic defined in the Prediction_Data input block instructions.

[7] If the reference value equals 0.0, the scaled sensitivities are set to zero.

Table 16. Contents of data-exchange files for analysis of model fit.
[Number of data lines = number of observations + number of prior information equations, except as follows. The _gm file has one line for each observation. The _pr file has one line for each prior information equation. The _xyztwr file has one line for each observation.]

File[1] Extension	Brief Description	"Column Tags" – Surrounded by double quotes in header			
_gm	Defines observation and prediction groups	GROUP NAME	MEMBER NAME	PLOT SYMBOL	
_nm	Normality of weighted residuals	WEIGHTED RESIDUAL	STANDARD NORMAL STATISTIC	PLOT SYMBOL	OBSERVATION or PRIOR NAME
_os	Observed and Simulated values	SIMULATED EQUIVALENT	OBSERVED or PRIOR VALUE	PLOT SYMBOL	OBSERVATION or PRIOR NAME
_pr	Prior information	PRIOR NAME	PLOT SYMBOL	EQUATION	NATIVE SPACE PRIOR VALUE
_r	Residuals	RESIDUAL	PLOT SYMBOL	OBSERVATION or PRIOR NAME	
_w	Weighted residuals	WEIGHTED RESIDUAL	PLOT SYMBOL	OBSERVATION or PRIOR NAME	
_ws	Weighted residual, Simulated equivalent	SIMULATED EQUIVALENT	WEIGHTED RESIDUAL	PLOT SYMBOL	OBSERVATION or PRIOR NAME
_ww	Weighted observed, Weighted simulated	WEIGHTED SIMULATED EQUIVALENT	WEIGHTED OBSERVED or PRIOR VALUE	PLOT SYMBOL	OBSERVATION or PRIOR NAME
_xyztwr	Coordinates, residuals, weighted residuals	X Y Z T	WEIGHTED RESIDUAL	RESIDUAL	PLOT SYMBOL / OBSERVATION NAME

[1]files _wt and _wtpri are listed in table 17 rather than here because they have a compressed matrix format as described in Chapter 10.

151

Table 17. Contents of the data-exchange file with extensions _wt and _wtpri, which contain the weighting for observations and prior information, respectively. Each file contains a weight matrix and the square-root of a weight matrix. [The matrices are stored in the compressed format described in Chapter 10.]

Sequentially listed lines	Data
One line	The number of matrices in the file. The number equals 2 for UCODE_2005.
One line	Label: COMPRESSEDMATRIX
One line	Three numbers: Number of non-zero values, the number of rows in the weight matrix[1], and the second number repeated.
One line for each non-zero element of the weight matrix	Two numbers on each line, an integer followed by a real number, which equal the matrix position and a non-zero element of the matrix.
One line	Label: COMPRESSEDMATRIX
One line	Three numbers: Number of non-zero values, the number of rows in the weight matrix[1], and the second number repeated.
One line for each non-zero element of the square-root of the weight matrix	Two numbers on each line, an integer followed by a real number: Matrix position, non-zero element of the square-root of the weight matrix

[1] For the _wt file, the number of rows equals the number of observations. For the _wtpri file, the number of rows equals the number or prior information equations.

Example _wt data-exchange file including a diagonal weight matrix and its square root. The weight matrix has diagonal terms equal to 4.0, 9.0, and 1.0; the square-root of the weight matrix has diagonal terms equal to 2.0, 3.0, and 1.0.

```
2
COMPRESSED MATRIX
3   3   3
1   4.0
5   9.0
9   1.0
COMPRESSED MATRIX
3   3   3
1   2.0
5   3.0
9   1.0
```

Table 18. Contents of the sensitivity analysis data-exchange files, ordered from most to least commonly used to construct graphs or tables.

[DSS, dimensionless scaled sensitivities; %, percent. Number of data rows equals number of observations, except as follows: for _sc, the number of data rows equals the number of parameters; for _pcc, the number of data rows equals the number of parameter pairs with correlations equal to or greater than 0.85. Numbers are added to the end of ParamName to emphasize when they are listed in order.]

File Extension	Brief Description	Column Tags Surrounded by double quotes in header. Capitalized COLUMN TAGS are used literally. ParamName is replaced by user defined names.			
_sc[1]	Composite scaled sensitivities	PARAMETER NAME	COMPOSITE SCALED SENSITIVITY	LARGEST DSS FOR THE PARAMETER	OBSERVATION WITH LARGEST DSS	Two more columns with second largest DSS
_so	Leverage	OBSERVATION or PRIOR NAME	PLOT SYMBOL	LEVERAGE	LARGEST DSS	PARAMETER WITH LARGEST DSS
_sd[1]	Dimensionless scaled sensitivities	OBSERVATION	PLOT SYMBOL	ParamName1	ParamName2	Total number of columns: number of parameters +2
_pcc	Parameter correlation coefficients	ParamName	ParamName	CORRELATION		
_s1[1]	One percent scaled sensitivities	OBSERVATION NAME	PLOT SYMBOL	ParamName1	ParamName2	Total number of columns: number of parameters +2
_su[1]	Unscaled sensitivities	OBSERVATION NAME	PLOT SYMBOL	ParamName1	ParamName2	Total number of columns: number of parameters +2
_supri[1,2]	Unscaled sensitivities	PRIOR NAME	PLOT SYMBOL	ParamName1	ParamName2	Total number of columns: number of parameters +2

[1] Values for up to 500 parameters are printed as a set using long lines in the output file. Additional parameters are printed in subsequent additional sets in the same file. Each set has the headers shown and the observation name and plot symbol in the first two columns.

[2] Only produced when prior information is defined.

Table 19. Contents of parameter-analysis data-exchange files.
[iteration, parameter-estimation iteration; #, a positive integer is printed; NP, the number of adjustable or estimated parameters. Numbers are added to the end of ParamName and ObsName to emphasize when they are listed in order.]

File Extension	Description	Column Tags Surrounded by double quotes in header. Capitalized COLUMN TAGS are used literally. ParamName and Obsname are replaced by user defined names.			
_pa	Parameter values, in columns.	`PARAMETER:ParamName` `ITERATION  ESTIMATE`	Repeated for each parameter.		
_pasub	Parameter values for substitution	`ITERATION: #` `PARAMETER  ESTIMATE`	Repeated for each iteration. Formatted for the Parameter_Values input block.		
_paopt	All defined parameters	`PARAMETER  OPTIMAL` `VALUE`	`LOG TRANSFORM` `Native= 0  Log=1`	PINC[2]	
_pc	Estimated parameters	Fourteen column tags [3]			
_ss	Sum-of-squared weighted residuals	`ITERATION`	`SSWR-` `(OBSERVA-` `TIONS` `ONLY)`	`SSWR-` `(PRIOR` `INFORMA-` `TION ONLY)`	`SSWR-` `(TOTAL)`	`# OBSERVA-` `TIONS` `INCLUDED`
_sos [1]	Objective-function values	`SUM-OF-` `SQUARED` `WEIGHTED` `RESIDUALS`	`Param-` `Name1`	`Param-` `Name2`	`Param-` `Name3`	Number of columns = NP+1.
_b1 [1]	Parameter sets for _b2	`ParamName1 ParamName2 ParamName3 ..` Number of data rows = 2 × number of estimated parameters			
_b2 [1]	From Model Linearity Mode	`ObsName1   ObsName2   ...` Number of data rows = 2 × number of estimated parameters			
_mc [1]	Parameter correlation matrix	`ParamName1 ParamName2 ParamName3 ...` Row 1 of parameter correlation matrix Row 2 ...			The matrix has NP rows and NP columns
_mv [1]	Parameter covariance matrix	`ParamName1 ParamName2 ParamName3 ...` Row 1 of parameter variance-covariance matrix Row 2 ...			The matrix has NP rows and NP columns

[1] Up to 500 parameter values or simulated values are printed as a set using long lines in the output file. Additional values are printed in subsequent additional sets in the same file. Each set has the headers shown and the observation name and plot symbol in the first two columns.

[2] Options for PINC: -1, adjustable=no; 0, not estimated because of dynamic omission of insensitive parameters or imposed constraints; 1, estimated.

[3] Column Tags: (1) PARAMETER NAME, (2) OPTIMAL VALUE, (3) LOWER LIMIT (NATIVE), (4) UPPER LIMIT (NATIVE), (5) LOG TRANSFORM Native= 0 Log=1, (6) OPTIMAL VALUE (REGRESSION) (7) STANDARD DEVIATION (REGRESSION) (8) COEFFICIENT OF VARIATION (REGRESSION) (9) STANDARD DEVIATION (NATIVE) (10) COEFFICIENT OF VARIATION (NATIVE) (11) REASONABLE RANGE MINIMUM, (12) REASONABLE RANGE MAXIMUM (13) IN REASONABLE RANGE? (Yes=1, No=0), (14) CONFIDENCE INTERVAL INCLUDES REASONABLE

VALUES? (Yes=1 No=0). The values in columns 3 and 4 are the lower and upper limit of a linear individual 95-percent confidence interval. For log-transformed parameters, item 9 is calculated as $s_b{}^2 = \exp[2.3(s_{\log b})^2 + 2.0 \times \log b][\exp(2.3(s_{\log b})^2) - 1.0]$ where b is item 2 and $s_{\log b}$ is item 7.

Table 20. Contents of prediction analysis files produced by UCODE_2005.
[PSD, Parameter standard deviation; RefValue, Reference value; PredValue, Predicted value; Param, Parameter Value. Numbers are added to the end of ParamName and ObsName to emphasize when they are listed in order.]]

File Extension	Contents				
UCODE_2005 main output files for prediction runs					
#upred	Summary of UCODE_2005 run with Prediction=yes in the UCODE_Control_Data input block.				
Description	Column Tags Surrounded by double quotes in header. Capitalized COLUMN TAGS are used literally. ParamName and Obsname are replaced by user defined names.				
Prediction data-exchange files produced by UCODE_2005. Can be used in many ways, including as input files for LINEAR_UNCERTAINTY					
_gmp	Prediction groups	GROUP NAME	MEMBER NAME	PLOT SYMBOL	
_p	Predictions	PREDICTED VALUE	PLOT SYMBOL	PREDIC-TION NAME	
_pv	Prediction variances	PREDIC-TION VARIANCE	PREDIC-TION NAME	PREDIC-TION GROUP	
_spu[1,2] _spsr[1,2,3] _spsp[1,2,3] _sppr[1,2,3] _sppp[1,2,3]	Scaling: [unscaled] [×PSD/RefValue] [×PSD/PredValue] [×Param/RefValue] [×Param/PredValue]	PREDIC-TION NAME	PLOT SYMBOL	Param-Name1	Param-Name2 Number of columns is number of parame-ters +2

[1] These data-exchange files contain prediction scaled sensitivities and differ only in how the sensitivities are scaled. The first two letters of the file extensions are an s for sensitivity and a p for prediction. The third letter is a u for unscaled, an s if the parameter standard deviation is used in the scaling, and a p if the parameter value is used in the scaling. When present, the fourth letter is r if the reference value is used in the scaling and p if the predicted value is used in the scaling. The scaling is described in the brackets of the second column of the table.

[2] Values for up to 500 parameters are printed as a set using long lines in the output file. Additional parameters are printed in subsequent additional sets in the same file. Each set has the headers shown and the observation name and plot symbol in the first two columns.

[3] If the scaling in the denominator equals 0.0, the scaled sensitivities are set to zero.

Table 21. Format of data-exchange files with basic data from the model (_dm, _dm_init and, for predictions, _dmp). Each line is composed of a label followed by data.

Label	Example Data
_dm	
"MODEL NAME:"	"ex1fullprior"
"MODEL LENGTH UNITS:"	"M"
"MODEL MASS UNITS:"	"NA"
"MODEL TIME UNITS:"	"D"
"NUMBER ESTIMATED PARAMETERS:" [1]	5
"ORIGINAL NUMBER ESTIMATED PARAMETERS:"[1]	6
"TOTAL NUMBER PARAMETERS:" [1]	9
"NUMBER OBSERVATIONS INCLUDED:" [2]	35
"NUMBER OBSERVATIONS PROVIDED:" [2]	35
"NUMBER PRIOR:"	3
"CALCULATED ERROR VARIANCE (FINAL):" [3]	1.74556582272
"STANDARD ERROR OF THE REGRESSION:" [3]	1.32119863106
"MAXIMUM LIKELIHOOD OBJECTIVE FUNCTION (MLOF):"	-29.52974
"AIC (MLOF + AIC PENALTY): "	-17.65474
"BIC (MLOF + BIC PENALTY): "	-11.34181
"HQ (MLOF + HQ PENALTY): "	-16.61653
"KASHYAP (MLOF + KASHYAP PENALTY): "	11.96280
"LOG DETERMINANT OF FISHER INFORMATION MATRIX:"	32.49399
"RN2 DEPENDENTS:"	0.9756259
"RN2 DEPENDENTS AND PRIOR:"	0.9783053
"NUMBER OF ITERATIONS"	4
_dm_init	
"CALCULATED ERROR VARIANCE (INITIAL): "	100002953.094
_dmp	
"NUMBER OF PREDICTION GROUPS: "	2

[1] "NUMBER ESTIMATED PARAMETERS" is less than "ORIGINAL NUMBER ESTIMATED PARAMETERS" if adjustable parameters are excluded from the regression during the parameter-estimation iteration. This can occur because of dynamic omission of insensitive parameters or the imposition of constraints, as controlled by keywords on the input blocks used to define parameters. "ORIGINAL NUMBER ESTIMATED PARAMETERS" is less than "TOTAL NUMBER PARAMETERS" if for any parameters the keyword Adjustable=no.

[2] "NUMBER OBSERVATIONS INCLUDED" is less than "NUMBER OBSERVATIONS PROVIDED" if simulated values could not be obtained for some observations.

[3] If regression does not converge and reaches the maximum number of iterations, these are assigned the value 0.100000000000E+31.

Chapter 15: EVALUATION OF RESIDUALS, NONLINEARITY, AND UNCERTAINTY

Thorough analysis of a calibrated model requires that the match of the simulated values to the observations be evaluated and presented. In addition, often it is useful to evaluate the relative dominance of the different observations in parameter estimation. Finally, when model predictions are to be used for resource management, remediation planning, and so on, the uncertainty of the predictions needs to be communicated along with the predictions themselves. To address these issues, six codes are provided as part of UCODE_2005; three are presented in this chapter and another three are presented in Chapter 17. Their use is described in this chapter, Chapters 16 and 17, and in Hill and Tiedeman (in press).

This chapter discusses RESIDUAL_ANALYSIS, LINEAR_UNCERTAINTY, AND MODEL_LINEARITY. Flowcharts showing how these codes coordinate with runs of UCODE_2005 are shown in figure 16. The descriptions provided here include short statements of the purpose of the code, descriptions of the input files, and a listing of the steps that need to be followed to execute the program. All of the required input files are produced by UCODE_2005; instructions for optional user-prepared input files are provided. Use of the output files produced by UCODE_2005, including the three codes described in this chapter, is discussed in Chapter 16.

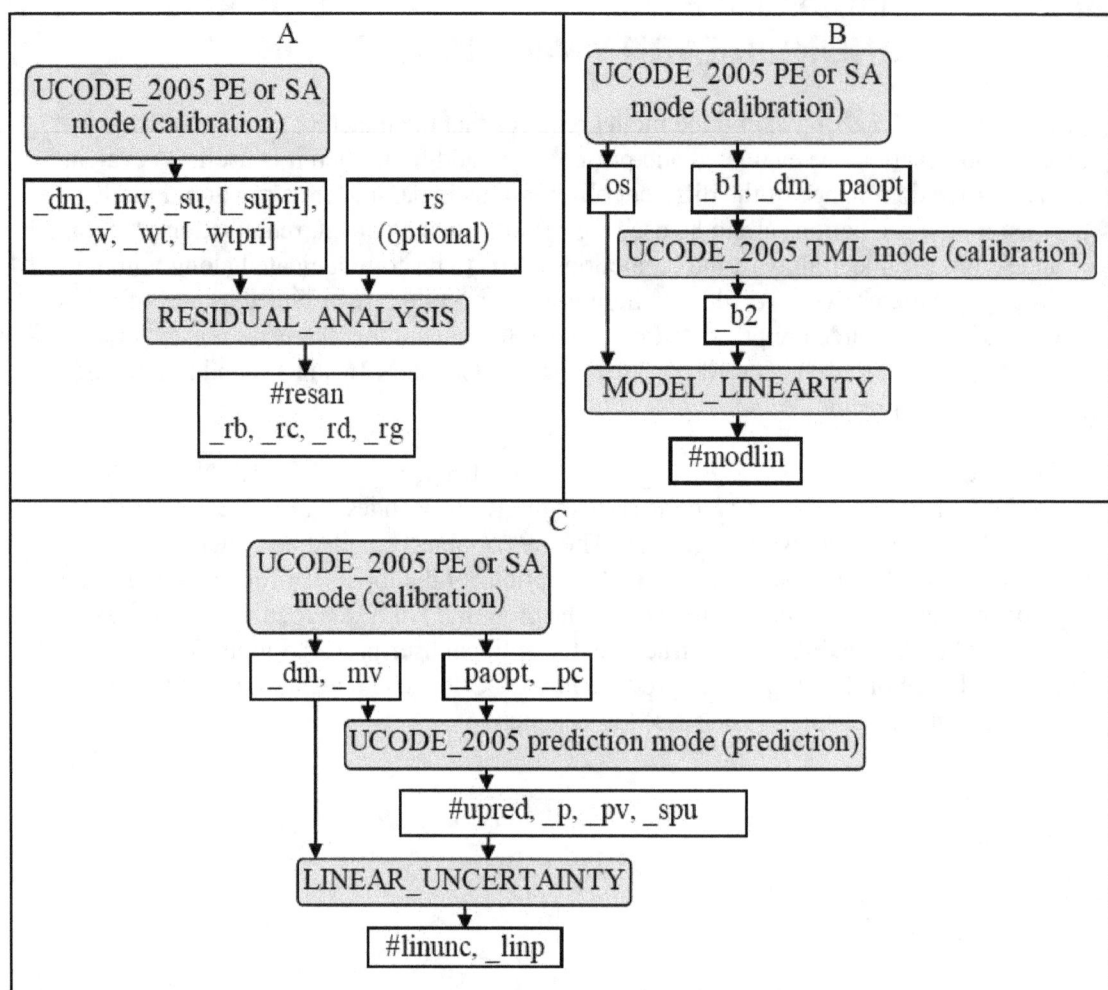

Figure 16. Flow charts with UCODE_2005 runs for (A) RESIDUAL_ANALYSIS, (B) MODEL_LINEARITY, and (C) LINEAR_UNCERTAINTY. Shaded boxes identify code executions. For the UCODE_2005 runs, only output files needed by the included codes are listed. UCODE_2005 modes are indicated as PE, parameter-estimation; SA sensitivity-analysis; and TML, test-model-linearity. (calibration) indicates that calibration conditions need to be simulated; (prediction) indicates that prediction conditions need to be simulated.

RESIDUAL_ANALYSIS: Test Weighted Residuals and Identify Influential Observations

RESIDUAL_ANALYSIS performs two functions that are described in the following paragraphs. RESIDUAL_ANALYSIS is the most commonly used of the post-processing programs. Often it is included in the script, batch file, or macro being used to execute UCODE_2005 so that the results are always available.

The first function performed by RESIDUAL_ANALYSIS is to test the weighted residuals for acceptable deviations from being independent (lacking any correlation) and normally distributed, as suggested by Draper and Smith (1998), Cooley and Naff (1990), Hill (1998, p. 24), and Hill and Tiedeman (in press, Chapter 6.3.6). Deviations are characterized using normal probability graphs of the weighted residuals (produced using the _nm file of table 16) and of generated random numbers (produced using the files with extensions _rd and _rg generated by RESIDUAL_ANALYSIS). The generated numbers are from a normal distribution with a mean of 0.0. Two types of generated random numbers are considered: (1) independent with a variance of 1.0 and (2) correlated as expected for the weighted residuals considering the regression performed. Correlated weighted residuals can result from the fitting process of the regression.

The weighted-residual test needs to be conducted if the normal probability graph of the weighted residuals, produced using the _nm file, does not approximate a straight line. Greater deviation from a straight line indicates a greater chance that the weighted residuals cannot be considered as random and normally distributed. A statistic, R_N^2 (Hill and Tiedeman, in press, Chapter 6.3.5, with critical values in their Appendix D) is printed in the fn.#uout file. Values of R_N^2 that are too much less than 1.0 indicate that the weighted residuals are less likely to be independent and normally distributed. A message printed in the fn.#uout file compares the calculated value of R_N^2 to the appropriate critical values and states the conclusion to be drawn from this comparison. To test the weighted residuals, RESIDUAL_ANALYSIS needs to be executed only if the weighted residuals deviate significantly from being normally and independently distributed, as indicated by small values of R_N^2 and normal probability graphs (_nm) for which the points do not fall on a straight line.

The second function of RESIDUAL_ANALYSIS is to calculate statistics that can be used to identify observations that are influential in the regression. The statistics calculated are Cook's D (fn._rc) and DFBetas (fn._rb), which are described by Belsley and others (1980) and Cook and Weisberg (1982), and applied to the development of a ground-water model by Yager (1998).

To produce the input files used by RESIDUAL_ANALYSIS, UCODE_2005 needs to be run in the sensitivity-analysis or parameter-estimation mode (Table 3; Fig. 16A) with DataExchange=yes in the UCODE_Control_Data input block. As noted in Table 3, sensitivity-analysis mode requires that SenMethod = -1, 0, or 2. The underscore file contents are described in tables 17 through 20 and table 22.

Chapter 15. Evaluation of Residuals, Nonlinearity, and Uncertainty
-- RESIDUAL ANALYSIS--

RESIDUAL_ANALYSIS is then executed using the same filename prefix, fn, on the command line used by UCODE_2005, which it will use to find the fn._xx files. For example, the following command would find the executable up two directories and down in the bin directory, and execute residual_analysis.exe using underscore file with the filename prefix "ex1":

..\..\bin\residual_analysis ex1

More information about executing RESIDUAL_ANALYSIS is provided in Chapter 4.

An optional input file can be created to override the RESIDUAL_ANALYSIS defaults. The file needs to be stored in the directory with the UCODE_2005 underscore files and to be named fn.rs, where fn is the filename prefix defined in the run command. RESIDUAL_ANALYSIS checks to see if such a file is present, and if so uses the items specified by the user in that file.

The default number of sets of random numbers is four, which should be sufficient in most circumstances. Additional sets are occasionally needed to conclusively test a set of residuals. In such circumstances, an fn.rs file needs to be created before executing RESIDUAL_ANALYSIS.

Change the random number seed to generate a different set of random numbers.

Chapter 15. Evaluation of Residuals, Nonlinearity, and Uncertainty
-- RESIDUAL ANALYSIS--

Table 22. Keywords that can occur in the optional RESIDUAL_ANALYSIS input file fn.rs, where fn is the filename prefix defined on the command line of UCODE_2005 and RESIDUAL_ANALYSIS.

[The keywords are placed in the RESIDUAL_ANALYSIS_Control_Data input block. The input block format is defined in Chapter 5.]

Keyword	Description	Default	Options
NSETS	Number of sets of random numbers to be generated	4	A positive integer.[1]
SEED	Seed used to generate random numbers	104857	A positive integer.[2]
CALC_RANDOMNUMBERS	Calculate random numbers. Create fn._rd and fn._rg files	Yes	Yes/No
CALC_COOKSD	Calculate Cook's D statistics for each observation, print summary in fn.#rs file, and create fn._rc file	Yes	Yes/No
CALC_DFBETAS	Calculate DFBetas statistics for each observation for each parameter. Print summary in fn.#rs file. Create fn._rb file	Yes	Yes/No
The following keywords control printing to the fn.#rs output file. Specify Yes to print the data described; No not to print the data. These keywords do not affect production of the data-exchange files listed in Table 23.			
PRINT_PAR_VAR_COV_MATRIX	Print parameter variance-covariance matrix	No	No/Yes
PRINT_SQRT_WT	Print the square root of the weight matrix as a second matrix	No	No/Yes
PRINT_UNSCALED_SENS	Print unscaled sensitivities calculated for the optimized parameter values	No	No/Yes
PRINT_COOKSD	Print Cook's D statistics for each observation	No	No/Yes
PRINT_RB	Print DFBetas statistics for each observation for each parameter	No	No/Yes
PRINT_RD	Print the sets of random numbers, if NSETS is greater than 0	No	No/Yes
PRINT_RG	Print the sets of correlated random numbers, if NSETS is greater than 0	No	No/Yes
PRINT_RES_VAR_COV_MATRIX	Print residual variance-covariance matrix	No	No/Yes
PRINT_RES_CORRELATION_MATRIX	Print residual correlation matrix	No	No/Yes

[1] Commonly between 4 and 10. If NSETS=0, no sets are generated.
[2] Needs to be between 1 and 1,048,575.

Chapter 15. Evaluation of Residuals, Nonlinearity, and Uncertainty
-- RESIDUAL ANALYSIS--

Example of the optional *fn*.rs file:

```
# Omitted keywords use default values.
# Keywords are case insensitive
Begin RESIDUAL ANALYSIS Control Data
  NSETS=5
  PRINT RD=True
  Print rg=true
End RESIDUAL_ANALYSIS_Control_Data
```

Table 23. Brief description of RESIDUAL_ANALYSIS input and output files.

File Extension	Brief description	Analyzed quantity
RESIDUAL_ANALYSIS input files produced by UCODE_2005 (content described with UCODE_2005 output files)		
_dm	Model name, units, and summary statistics	Model fit
_mv	Parameter variance-covariance matrix	Parameter
_su	Unscaled sensitivities for observations	Sensitivity
_supri	Unscaled sensitivities for prior information	Sensitivity
_w	Weighted residuals for observations and prior information, and scaled differences between predictions and reference values	Model fit
_wt	Weights for observations	--
_wtpri	Weights for prior information	--
RESIDUAL_ANALYSIS input file created by the user (optional)		
.rs	Number of sets of random numbers, seed for random number generator, output control variables	
RESIDUAL_ANALYSIS output files		
#rs	Runtime information for RESIDUAL_ANALYSIS. This file is useful only for debugging.	--
_rb	DFBetas statistics for each observation for each parameter	Model fit; Sensitivity
_rc	Cook's D statistic for each observation	Model fit; Sensitivity
_rd	Ordered uncorrelated numbers vs. probability plotting positions for each observation.	Model fit
_rg	Ordered correlated numbers vs. probability plotting positions for each observation.	Model fit

Table 24. Contents of RESIDUAL_ANALYSIS output files.
[ParamName is replaced by user-defined names. Numbers are added to the end of ParamName to emphasize when they are listed in order.]

File Exten-sion	Description				
#resan	Summary of a RESIDUAL_ANALYSIS run				
	Description	**"Column Tags" – Surrounded by double quotes in header**			
Data-Exchange files					
_rb[1]	DFBetas for each parameter for each observation. Extreme values are listed in the #rs file.	OBSERVA-TION OR PRIOR INFOR-MATION NAME	PLOT SYMBOL	ParamName1 ParamName2 ...	
_rc	Cook's D for each observation. Observations with extreme values are listed in the #rs file.	COOK'S D	OBSERVA-TION OR PRIOR INFOR-MATION NAME	PLOT SYMBOL	
_rd	Uncorrelated normal random numbers.	ORDERED INDE-PENDENT DEVIATE	STANDARD NORMAL STATIS-TIC	PLOT SYMBOL	OBSERVA-TION or PRIOR INFOR-MATION NAME	RANDOM NUMBER SET NO.[2]
_rg	Correlated normal random numbers.	ORDERED CORRE-LATED DEVIATE	STAN-DARD NORMAL STATIS-TIC	PLOT SYMBOL	OBSERVA-TION or PRIOR INFOR-MATION NAME	RANDOM NUMBER SET NO.[2]

[1] Values for up to 500 parameters are printed as a set using long lines in the output file. Additional parameters are printed in subsequent additional sets in the same file. Each set has the headers shown and the observation name and plot symbol in the first two columns.
[2] The header line is repeated for each set of random numbers.

LINEAR_UNCERTAINTY: Calculate Linear Confidence and Prediction Intervals on Predictions Simulated with Estimated Parameter Values

Predictions produced by a calibrated model are much more useful when they are reported with an evaluation of prediction uncertainty. Use of regression in model calibration, as supported by UCODE_2005, provides clear methods by which measures of the parameter uncertainty can be propagated into measures of prediction uncertainty. The parameters included can be those estimated by regression as well as those which, because of insensitivity, parameter correlation, or both, could not be estimated by regression. Inclusion of the unestimated parameters in the evaluation of prediction uncertainty was mentioned in Chapter 3 and is discussed further by Hill and Tiedeman (in press). An example is provided in Appendix C. The uncertainty in predictions is usually represented using confidence and prediction intervals, which were defined in Chapter 3.

LINEAR_UNCERTAINTY calculates 95-percent linear confidence and prediction intervals on predictions. Two issues of concern when using these intervals are as follows. First, the uncertainty propagated from parameters generally underestimates the total prediction uncertainty because it does not fully account for what is often called conceptual model error. Second, the linearity of the method may produce intervals that are too small or too large for predictions simulated with nonlinear models. These issues are discussed briefly in the following paragraphs.

The uncertainty associated with alternative models can be incorporated into the analysis with methods such as those used in the computer program MMRI discussed by Poeter and Anderson (in press) and documented by Poeter, Hill, and Banta (USGS, written commun., 2005). In MMRI, intervals calculated using different conceptual models are integrated to create an interval for each prediction that reflects the conceptual model uncertainty. The alternative models can be developed based on deterministic or stochastic arguments.

Conceptual model error is likely to be smaller if the model is designed such that the defined parameters include those aspects of the system that are poorly constrained by the observations but are important to predictions. Suggestions for defining parameters are discussed in guideline 3 of Hill and Tiedeman (in press) and Hill (1998).

The use of linear intervals to quantify model uncertainty and the advantages and disadvantages of nonlinear intervals are analyzed by Christensen and Cooley (1999), Cooley (2004), and Christensen and Cooley (2005), and discussed by Hill (1998, p. 29-31), Hill and Tiedeman (in press, Chapter 8.3), and in Chapters 3, 16, and 17 of this report.

LINEAR_UNCERTAINTY calculates linear individual and simultaneous intervals, which were discussed in Chapter 3. For simultaneous intervals, all predictions to be considered simultaneously need to be produced in a single execution of

LINEAR_UNCERTAINTY. This is needed because the number of predictions being considered affects the critical value used to calculate the simultaneous intervals.

The predictions and their sensitivities needed by LINEAR_UNCERTAINTY are calculated using the prediction mode of UCODE_2005, generally using a model calibrated with the parameter-estimation mode (Table 3). The process model run(s) used to generate the predictions may simulate, for example, potential future pumpage, a climate-change scenario, and so on.

LINEAR_UNCERTAINTY only requires input files that are produced by two runs of UCODE_2005 (Fig. 16C). There are no optional, user provided input files.

To obtain linear measures of uncertainty, first, of course, the model to be used to calculate predictions needs to be developed. If this is achieved using the parameter-estimation capabilities of UCODE_2005 and no parameter modifications are needed, steps 1 and 2 below can be skipped.

Parameter modifications are needed when calculating confidence and prediction intervals if, during calibration, the following occur. (a) One or more parameters are held constant because

(i) Adjustable=no was specified in the Parameter_Data input block or

(ii) They were insensitive and dynamic omission of insensitive parameters was activated using the OmitInsensitive keyword of the Reg_GN_Controls input block or

(iii) The parameter values varied such that constraints were applied, as controlled by the Constrain keyword in the Parameter_Groups or Parameter_Data input block.

(b) One or more parameters were assigned prior information with smaller statistics than supportable by independent measurements. While these methods can be valid ways to obtain estimated parameter values during model calibration, it is important that the actual uncertainty in the parameters be included in the calculation of confidence and prediction intervals (Hill and Tiedeman, in press, Chapter 8.1; Hill, 1998, p. 25). Parameter modifications also can be needed to include parameters important to predictions that were not important to model calibration. For example, in Appendix C the advective-transport prediction is sensitive to porosity, which was irrelevant to model calibration with hydraulic heads and flow observations. To modify defined parameters, use steps 1 and 2.

3. If any parameters were held constant or given unrealistic prior information during regression, or if parameters are added, do the following:

    a. Consider creating a new directory because the run in step 2 will overwrite the files created by the parameter-estimation mode.

    b. Change the parameter definitions as needed. Possibilities include:

        i. Activate parameters that were held constant during calibration and apply appropriately weighted prior information to them as warranted by independently available data.

      ii. For estimated parameters assigned prior information with statistics that were smaller than supportable by independent data, assign statistics that are consistent with the independent data.

      iii. Add parameters important to predictions that were not included in model calibration. Add prior information as warranted by independent data. (An example is presented in Appendix C)

  c. Update the parameter values in the Parameter_Data input block or use the Parameter_Values input block. Estimated parameters need to be assigned the optimal values achieved through regression.

4. Execute UCODE_2005 in the sensitivity-analysis mode with SenMethod=-1, 0, or 2 (Table 3).

To simulate predictions, follow steps 3 through 5.

5. Prediction conditions often are different than calibration conditions. For example, changes in pumpage and changes in areal recharge caused by climate change can be imposed. The prediction conditions are imposed through input files of the process model(s). It is often useful for the calibration and prediction simulations to be placed in separate directories, as in the example files distributed with UCODE_2005 for the problem in Appendix C. Uncertainty in parameters characterizing such stresses can be included in the calculation of confidence and prediction intervals. For example, the uncertainty in areal recharge caused by climate change can be included by defining one or more parameters used to calculate the areal recharge and then assigning prior information that expresses the uncertainty in the parameter values. If differences are of concern, two model runs can be defined – one for the base case and one for the predictive conditions. Equations can be used in the Prediction_Data or Derived_Prediction input blocks to calculate predicted differences of interest.

6. Execute UCODE_2005 in prediction mode with Prediction=yes. Output is in fn.#upred. Data-exchange files with file extensions _p and _spu are produced for use by LINEAR_UNCERTAINTY. Generally these input files are not accessed by the user. Prediction scaled sensitivities are printed in the following data-exchange files: _spsr, _spsp, _sppr, _sppp (table 21).

7. Execute LINEAR_UNCERTAINTY as discussed in Chapter 4 to calculate linear confidence intervals for the defined predictions. Output is to files with extensions #linunc, and _linp (table 26).

Chapter 15. Evaluation of Residuals, Nonlinearity, and Uncertainty
-- LINEAR_UNCERTAINTY--

Table 25. Brief description of input and output files for predictions.
[Files produced by UCODE_2005 are described in Table 20.]

File Extension	Brief description
UCODE_2005 prediction mode main output file	
#upred	Report from UCODE_2005 prediction mode run creating _p and _spu
LINEAR_UNCERTAINTY input files generated by UCODE_2005's Sensitivity Analysis or Parameter-Estimation mode	
_dm	Model data
_mv	Parameter variance-covariance matrix
LINEAR_UNCERTAINTY input files generated by UCODE_2005's Prediction mode	
_p	Predicted values
_spu	Unscaled sensitivities for predictions
LINEAR_UNCERTAINTY output files	
#linunc	Runtime information for LINEAR_UNCERTAINTY
_linp	Predicted values and their linear confidence intervals

Table 26. Contents of LINEAR_UNCERTAINTY output files.

File Extension	Graph / Information
#linunc	Summary of run generating _linp.

_linp	Six types of confidence intervals labeled as follows. Each label is in double quotes and is followed by columns of data with the labels listed below.
	`INDIVIDUAL 95% CONFIDENCE INTERVALS`
	`SIMULTANEOUS 95% CONFIDENCE INTERVALS`
	`UNDEFINED NUMBER OF SIMULTANEOUS 95% CONFIDENCE INTERVALS`
	`INDIVIDUAL 95% PREDICTION INTERVALS`
	`SIMULTANEOUS 95% PREDICTION INTERVALS`
	`UNDEFINED NUMBER OF SIMULTANEOUS 95% PREDICTION INTERVALS`

	Columns labels – Surrounded by double quotes

	`PREDIC-` `PREDICTED` `LOWER` `UPPER` `STANDARD` `PLOT`
	`TION` `VALUE` `LIMIT` `LIMIT` `DEVIATION` `SYMBOL`
	`NAME`

MODEL_LINEARITY: Test Model Linearity

The linear intervals produced by LINEAR_UNCERTAINTY can accurately reflect the uncertainty of the simulated values only if the model is sufficiently linear (Seber and Wild, 1989; Cooley and Naff, 1990; Hill, 1994; Hill, 1998, p. 31-32; Hill and Tiedeman, in press, Chapter 8.3.1). Model nonlinearity can be tested using the modified Beale's measure presented by Cooley and Naff (1990) and also discussed by Hill and Tiedeman (in press) and Hill (1994). Ground-water models are nearly always nonlinear with respect to estimated parameter values, as discussed in Chapter 2 of this report. Although the modified Gauss-Newton optimization method and the statistical methods calculated by UCODE_2005 are useful even for problems which are quite nonlinear, more stringent requirements on linearity are needed for the linear confidence and prediction intervals produced by LINEAR_UNCERTAINTY (discussed in the last section of this chapter) to adequately represent uncertainty. The modified Beale's measure indicates the possible severity of the problem.

The modified Beale's measure indicates nonlinearity of the confidence region of the parameters and does not directly measure nonlinearity of the confidence and prediction intervals. One consequence of this is that it can be misleading if the predictive quantities are substantially different from the observed quantities used in the regression, or if predictive ground-water flow conditions are substantially different than calibration conditions. The more recent combined intrinsic model nonlinearity measure produced by MODEL_LINEARITY_ADV and discussed in Chapter 17 addresses this problem, and may be used in place of the modified Beale's measure.

The modified Beale's measure is calculated using two data-exchange files produced by UCODE_2005 and the post-processing program MODEL_LINEARITY (Fig. 16B). The two data-exchange files are the _b1 and _b2 files of tables 2, 15 and 19; Table 19 describes the file contents. The main output file produced by MODEL_LINEARITY is called fn.#modlin. The input and output files are listed in Table 27. The modified Beale's measure is printed near the bottom of the fn.#modlin file along with critical values. This information can be used to determine whether the calibrated model is roughly linear, intermediate, or nonlinear, with respect to the observations used for model calibration. The rest of the information in the #modlin file can be used to detect which observations and parameters contribute most to the nonlinearity.

Table 27. Brief description of MODEL_LINEARITY input and output files.

File Extension	Content
MODEL_LINEARITY input files produced by UCODE_2005 (contents described with UCODE_2005 output files)	
_b1	Parameter sets for calculating Beale's measure of linearity
_b2	Simulated equivalents for Beale parameter sets of _b1
MODEL_LINEARITY output file	
#modlin	Results of Beale's analysis by MODEL_LINEARITY, using b1 and b2 from UCODE_2005

As for LINEAR_UNCERTAINTY, parameter modifications are needed when calculating the modified Beale's measure if, during calibration, the parameters were (a) held constant or (b) assigned prior information with smaller statistics than supportable by independent measurements. While these methods are valid ways to constrain the estimated parameter values sufficiently to attain a stable regression during model calibration, it is important that the uncertainty in the parameters considered when calculating linear intervals be considered when evaluating model linearity (Hill and Tiedeman, in press, Chapter 8.1; Hill, 1998, p. 25). If this was done for the LINEAR_UNCERTAINTY code, the _b1 file will have already been created. If not, use steps 1 to 2 described for the LINEAR_UNCERTAINTY code to modify defined parameters and then proceed with steps 3 and 4 listed here to evaluate model linearity.

3. Execute UCODE_2005 with linearity=yes. This will generate an _b2 file.

4. Execute MODEL_LINEARITY with the filename prefix of the UCODE_2005 files on the command line. This will generate the fn.#modlin file.

Chapter 16: USE OF OUTPUT FROM UCODE_2005, RESIDUAL_ANALYSIS, MODEL_LINEARITY, AND LINEAR_UNCERTAINTY

UCODE_2005 provides substantial flexibility in performance, as indicated by the modes listed in Table 3. A large number of output files can be produced, as shown in table B-1. The following sections describe how these files commonly are used given different modes of UCODE_2005 execution.

Output Files from UCODE_2005 Forward Mode

For a forward simulation, UCODE_2005 runs the process model(s) once using the specified parameter values. The main UCODE_2005 output file is fn.#uout. If DataExchange=yes in the UCODE_Control_Data input block, the graphical analysis files listed in Table 15 are all produced except for the _sos file. In addition, if prior information is used the _pr file is produced. The main output file fn.#uout needs to be used to check for errors in the forward simulation and the definition of observations.

The weighted residuals included in several of the files listed in Table 15 reflect (1) the model fit given the expected accuracy of the observations, (2) the existing model configuration, (3) the parameter values used, and (4) the procedure used to calculate the equivalent simulated values that are compared with the observations. Large discrepancies between simulated and observed values need to be investigated and may indicate, for example, that there is a data input error, or a conceptual error in the model configuration or in the calculation of the simulated values. Inspection of these values and correction of obvious problems can eliminate many hours of frustration. Use of the _ws file to graph weighted residuals and simulated values will clearly show whether there are large discrepancies between observed and simulated values. If there are large discrepancies, it is important to investigate whether they are caused by errors in the process-model input files, the input of observations, or in the production of the simulated values. It is essential for UCODE_2005 to perform correctly in this mode. Proceeding with errors will result in an invalid regression and wasted time.

Output Files from the UCODE_2005 Sensitivity-Analysis Mode

This mode is achieved as described in Table 3. The unique benefit of this mode is that scaled sensitivities and parameter correlation coefficients, which are the primary statistics needed for the sensitivity analysis described by Hill and Tiedeman (in press) and Hill (1998), are calculated using the starting parameter values. These statistics are calculated for parameters with keyword Adjustable=yes in the Parameter_Data input block, without proceeding on to a series of parameter-estimation iterations. The statistics are printed in the main output file if requested by user-specified flags in the UCODE_Control_Data input block.

The scaled sensitivities produced by the sensitivity-analysis mode are discussed in the preceding section. The parameter correlation coefficients that are produced can be used to identify highly correlated parameter pairs. The presence of highly correlated parameters can be problematic during parameter estimation because of the difficulty of determining unique values for highly correlated parameters. The utility of the parameter correlation coefficients depends on the accuracy of the sensitivities, as discussed by Hill and Østerby (2003). The parameter correlation coefficients calculated using more accurate sensitivity-equation sensitivities tend to be more useful. Parameter correlation coefficients produced using perturbation sensitivities are not as reliable. Parameter correlation coefficients can change substantially as the parameter values change, and should be reevaluated often during model calibration if non-unique parameters are suspected.

Maps from UCODE_2005 Sensitivity Analysis Mode

If SenMethod=0 or -1 in the Parameter_Data input block, the process model may produce files that can be used to produce maps of scaled sensitivities or adjoint states. Such maps are generally discussed in the documentation of the process models.

Though not often used quantitatively, these maps can be used to better understand the influence of different parameters on the calculation of simulated values such as hydraulic head in ground-water models. As the number of parameters, model layers, and time steps increases, the number of possible maps can be overwhelming, but judicious map production can produce important insights into system dynamics.

Tables of Scaled Sensitivities Produced for the UCODE_2005 Sensitivity-Analysis, Parameter-Estimation, and Prediction Modes

Depending on the designations of keywords StartSens, IntermedSens, FinalSens, and DataExchange in the UCODE_Control_Data input block, tables of dimensionless and composite scaled sensitivities and (or) one percent scaled sensitivities are printed in the main UCODE_2005 output file (fn.#uout) and in the _sc, _sd, and _s1 files (Table 15). For the parameter-estimation mode, in which nonlinear regression is performed, the main output file can include tables of scaled sensitivities calculated using the starting, intermediate, and final parameter values, or any combination. The _sc, _sd, and _s1 files always contain scaled sensitivities calculated using the final parameter values. The _so file summarizes the leverage of each observation on the regression. The use of dimensionless, composite, and one-percent scaled sensitivities, and leverage is discussed in Hill and Tiedeman (in press) and Hill (1998), and briefly summarized in the following paragraphs.

Dimensionless scaled sensitivities can be used to determine which observations are likely to be most important to the estimation of each parameter. They often do not, however, identify observations that reduce parameter correlation because these observations may not have large dimensionless scaled sensitivities. Bar charts can be used to readily indicate the observations with the largest dimensionless scaled sensitivities.

Chapter 16: Use of Output from UCODE_2005, RESIDUAL_ANALYSIS, MODEL_LINEARITY, and LINEAR_UNCERTAINTY

Composite scaled sensitivities can be used to evaluate whether the available observations are likely to provide enough information to allow estimation of defined parameters. Generally it is difficult to estimate parameters with composite scaled sensitivities that are more than a factor of 100 less than the maximum composite scaled sensitivity for a set of estimated parameters. Insensitive parameters can cause poor regression performance. UCODE_2005 can dynamically omit insensitive parameters as described for the Reg_GN_Controls input block of Chapter 6, or reduced execution times can be achieved by assigning insensitive parameters Adjustable=no as described in Chapter 7.

Composite scaled sensitivities are generally plotted using a bar chart. Plotting of such bar charts routinely during model calibration is important because both the nonlinearity of the sensitivities and the scaling will cause composite scaled sensitivities to change. These changes become important if they indicate that a parameter included in the estimation can no longer be supported by the observations, or a previously excluded parameter probably can be estimated given the updated version of the model, but experience suggests that such dramatic changes are rare. It is important to include both estimated and unestimated parameters in bar charts of composite scaled sensitivities when the modeling results are published.

Prediction scaled sensitivities are provided in the underscore files: spsr, spsp, sppr, sppp. The first of the four letters indicates that the file contains sensitivities. The second indicates that the sensitivities are for predictions. The third and fourth letters indicate how the sensitivities are scaled. The third letter is s for multiplication of the sensitivity by the parameter standard deviation or p for multiplication by the parameter value. The fourth letter is r for division by the reference value provided by the user, or p for division by the predicted value. The different scalings are provided because in different circumstances different scalings are useful. For example, meaningful reference values are not always available, so these scalings are not useful in some circumstances, while in other circumstances they are very useful. Prediction scaled sensitivities are often presented on bar charts, as shown in Tiedeman and others (2004) and Hill and Tiedeman (in press).

Output Files from the UCODE_2005 Parameter-Estimation Mode

UCODE_2005 performs nonlinear regression under the circumstances listed in Table 3. The main output file is fn.#uout. If keyword DataExchange=yes in the UCODE_Control_Data input block, many of the files of Table 15 are produced by UCODE_2005. If executed, RESIDUAL_ANALYSIS produces additional output files listed in Table 23. Often it is useful to set up batch files to execute RESIDUAL_ANALYSIS immediately after UCODE_2005 so that these files are routinely produced.

Four data-exchange files contain parameter values: _pa, _pasub, _paopt, and _pc (Table 19). Native parameter values are reported for log-transformed parameters.

The _pa file contains parameter values for each parameter-estimation iteration in a form suitable for plotting. This file can be used to produce graphs of parameter values for each iteration number using, for example, GW_CHART (Winston, 2000) or Microsoft Excel. Such graphs are useful in evaluating regression performance and problems.

The _pasub file contains parameter values from each parameter-estimation iteration in a format suitable for substitution into the Parameter_Values input block. These parameters are used to (1) investigate simulated equivalents to the observations and observations sensitivities calculated with parameter values from intermediate parameter-estimation iterations, (2) restart the regression using values from the final or intermediate parameter-estimation iterations, which is useful if the regression has deviated from reasonable values or if long run times limit how many parameter-estimation iterations are pursued in a single run.

The _paopt file contains information from the end of the run for all defined parameter values. This information includes the status of each parameter. This is important because these parameters with keyword Adjustable=yes may not be adjustable at the end of a UCODE_2005 parameter-estimation mode run because of imposed constraints or dynamic omission of insensitive parameters (see instructions for the Parameter_Data input block).

The _pc data-exchange file reports statistics about the final parameter estimates, including individual 95-percent linear confidence intervals, the standard deviations and coefficients of variation of the estimates, and an analysis of whether the estimate contradicts the reasonable range specified in the Parameter_Data input block. Two issues are of concern.

First, parameter values, standard deviations and coefficients of variation for both the native and regression parameters are reported (table 19). For log-transformed parameters, the native and regression values are different. Standard deviations for the native values are calculated from the parameter variances (the diagonal of the _mv data-exchange file) using an equation presented by Hill and Tiedeman (in press, Chapter 7.2.4) or Hill (1998, p. 27). The native and regression coefficients of variation are calculated as the ratio of the appropriate standard deviations and parameter values. To indicate how well the estimates are determined, the native coefficients of variation are most informative. The linear confidence intervals listed are for the native parameter values. For log-transformed parameters they plot symmetrically about the estimate on a log axis. On an arithmetic axis the interval is asymmetric about the estimate.

Second, the _pc file reports the results of comparing the estimates and their confidence intervals to reasonable ranges. Two questions are addressed. (1) Does the estimate fall within the reasonable range? (2) If not, does the confidence interval on the estimate include any reasonable values? As discussed by Hill and Tiedeman (in press, Guideline 10) and Hill (1998, Guideline 9), this analysis can be used to detect model error.

The main output file includes information about the regression and indicates whether or not the regression converged. The main output file lists the statistics described in tables 29 through 31 on convergence. If convergence is not reached and Stats_On_Nonconverge=no in the Reg_GN_Controls input block, then UCODE_2005 does not calculate more accurate sensitivity for the latest parameter updates and does not print the typical statistics for the final parameter values. A sample main output file from a regression is included in Appendix C of this report. The best way to become familiar with the file is to review that example and the comments in tables 29 through 31.

Residual analysis can be accomplished using the statistics listed in Table 28 and the files listed in table 31. Examples of the files with their contents labeled are shown in Appendix C. File names listed in table 31 with two letters in the extension include two columns of values and generally are used to create x-y plots. File names listed in table 31 with a single letter in the extension contain only one column of values and generally are used to create maps, temporal plots, or higher-dimensional images of residuals. Each line includes the information related to one observation or piece of prior information. In all files, each line lists the observation name or, for prior information, the prior information name. Each line also lists their plot symbols. Comments about how to use the generated graphs are presented in table 31. Additional discussion can be found in Hill and Tiedeman (in press), Hill (1998), and references cited therein.

The quality and progress of the regression is evaluated using the information summarized in table 30. The information is printed throughout the fn.#uout file and is summarized in tables at the end of that file. If progress is reasonable and convergence is achieved, information in table 31 is used to evaluate the optimal parameter values.

During most model calibrations, UCODE_2005 regression runs are executed many times as various aspects of the model are changed to test hypotheses about the system. During calibration and after calibration is completed, predictions can be calculated, linear confidence and prediction intervals can be calculated to provide an indication of the prediction uncertainty, and the linearity of the model can be evaluated. The model output related to these capabilities is described in subsequent sections on the post-processing programs LINEAR_UNCERTAINTY and MODEL_LINEARITY.

Table 28. Residuals and model-fit statistics printed in the main UCODE_2005 output files for Sensitivity Analysis and Parameter Estimation modes.
[See example output file in appendix C of this report.]

Statistic as labeled in the main UCODE_2005 output file[1]	Comments
Table of data, group name, simulated values, residuals, and weighted residuals	Residuals are calculated as observations or prior information minus the simulated values. Use this table or data-exchange files to evaluate model fit.
MAXIMUM/ MINIMUM WEIGHTED RESIDUAL	The maximum and minimum weighted residuals indicate where the worst fit occurs, and often reveals gross errors.
AVERAGE WEIGHTED RESIDUAL	An average weighted residual near zero is needed for an unbiased model fit (usually satisfied if regression converges).
# RESIDUALS >= 0. # RESIDUALS < 0.	The number of positive and negative residuals indicates whether the model fit is consistently high or low. Preferably, the two values are about equal.
NUMBER OF RUNS	Number of sequences of residuals with the same sign (+ or -). Too few or too many runs can indicate model bias. The related statistic is printed and interpreted. Hill and Tiedeman (in press, Chapter 6.3.4) explain the test.
The following are printed in the main UCODE_2005 output file for parameter-estimation mode.	
LEAST-SQUARES OBJ FUNC (OBS. ONLY) (W/PARAMETERS)	Weighted least-squares objective function value. Given randomly distributed residuals and the same observations and weight matrix, a lower value of the least-squares objective function indicates a closer model fit to the data.[2]
NUMBER OF INCLUDED OBSERVATIONS = # OF #	An observation is omitted if a simulated value can not be calculated. For example, dry cells can result in omitted observations.
CALCULATED ERROR VARIANCE	Given randomly distributed residuals, smaller values are desirable. For values less than 1.0: model fit to data is better than is consistent with the statistics used to weight observations and prior information; for values greater than 1.0: fit is worse. See Hill and Tiedeman (in press, Guideline 6).
STANDARD ERROR OF THE REGRESSION	The square root of the calculated error variance. For utility, see Hill and Tiedeman (in press, Guideline 6).
CORRELATION COEFFICIENT W/PARAMETERS	R of Hill and Tiedeman (in press, Chapter 6.3.3). Values below about 0.9 indicate poor model fit.[2]
MAX LIKE OBJ FUNC (MLOF) AIC HANNAN BIC KASHYAP	The maximum likelihood objective function, and four other criteria for evaluating appropriate models.[3] Given randomly distributed residuals, lower values indicate better model fit.
LOG DETERMINANT OF FISHER INFORMATION MATRIX	Measure of the information provided by observations and prior information for the parameters included in the analysis. Larger values: more information.
SMALLEST AND LARGEST WEIGHTED RESIDUALS	The five smallest and five largest weighted residuals are printed. Weighted residuals for observations and prior information are considered.

Statistic as labeled in the main UCODE_2005 output file[1]	Comments
`CORRELATION BETWEEN ORDERED WEIGHTED RESIDUALS AND NORMAL ORDER STATISTICS`	R_N^2 of Hill and Tiedeman (in press, Chapter 6.3.5, Appendix D). Values above critical value indicate independent, normal weighted residuals, and that the points listed in the _nm file are likely to fall on a straight line.[2]

[1]THIS FONT is used for labels taken directly from the output

[2]To allow detection of poor fit to one type of regression data, these statistics are calculated both for (a) the observations and (b) the observations and prior information.

[3]The statistics are calculated two ways: adding to MLOF and adding to $n[\log \sigma^2]$. They are discussed in the MMRI documentation of Poeter and others (written commun., 2005).

Table 29. Regression performance measures printed in the main UCODE_2005 output file for parameter-estimation mode.

[These measures are printed for each parameter-estimation iteration; see example file in Appendix C of this report]

Performance measure as labeled in the GLOBAL file [1]	Comments
MARQUARDT PARAMETER	Used as described in Hill and Tiedeman (in press). Non-zero values indicate an ill-conditioned problem.
MAX. FRAC. CHANGE OCCURRED FOR PAR. "PARNAME "	The parameter for which the maximum fractional change occurs. If the regression does not converge, the parameters listed here are likely to be contributing to the problem.
MAX. FRACTIONAL PARAMETER CHANGE	Maximum fractional change calculated for any parameter in the parameter-estimation iteration. Used to determine convergence of parameter estimation as described for keyword TolPar in the Reg_GN_Controls input block.
CONVERGENCE TOLERANCE FOR THIS PARAMETER	Input by user as TolPar in the Reg_GN_Controls input block, or as tolerance in the Parameter_Data block.
MAXIMUM FRACTIONAL CHANGE ALLOWED FOR THIS PARAMETER	Input by user as MaxChange in the Reg_GN_Controls input block.
One of the following two is printed	
(1) If the maximum fractional change is less than allowed, the following is printed: ADJUSTMENTS TO PARAMETER CHANGE VECTOR WERE NOT REQUIRED	
(2) When damping control is active, the following five items are printed:	
CONTROLLING PARAMETER :	Parameter that results in the smallest value of the damping parameter.
CHANGE CALCULATED FOR CONTROLLING PARAM: USED:	The fractional change calculated for the controlling parameter and the change that is used after damping is applied.
OSCILL. CONTROL FACTOR (1 is NO EFFECT) =	If the solution is oscillating further damping may be imposed.
DAMPING PARAMETER (RANGE 0 TO 1) =	Values less than 1.0 indicate that the maximum fractional parameter change exceeded the MaxChange value, or that oscillation control was active. (Hill and Tiedeman, in press, Chapter 5.1.1, Appendix A).
UPDATED ESTIMATED PARAMETERS	Parameter values calculated as a result of this parameter estimation iteration.

[1] THIS FONT is used for labels taken directly from the output

Table 30. Parameter statistics printed in the main UCODE_2005 output file for the Sensitivity-Analysis and Parameter-Estimation modes.
[See example file in Appendix C. %, percent.]

Parameter statistic or characteristic[1]	Function of item in interpreting results
[2] DIMENSIONLESS SCALED SENSITIVITIES (SCALED BY B*(WT**.5)) (_sd, _so)	Indicates the importance of an observation to the estimation of a parameter or, conversely, the sensitivity of the simulated equivalent of the observation to the parameter. These values are listed in a table with a row for each observation and a column for each parameter.
[2] COMPOSITE SCALED SENSITIVITIES ((SUM OF THE SQUARED VALUES)/ND)**.5 (_sc)	Indicates the information content of all observations toward estimation of a parameter. Values less than 1.0 indicate little information. Values less than 0.01 times the largest value indicate parameters with relatively little information. For such parameters, regression is likely to have trouble converging, and estimating that parameter value will be difficult with the available observations.
ONE-PERCENT SCALED SENSITIVITIES (SCALED BY B/100) (_sl)	These scaled sensitivities have the dimensions of the observations, which can sometimes be useful.
Parameter covariance matrix (_mv)	Diagonal contains variances; off-diagonal terms are covariances. These values are used to calculate the statistics listed below.
Parameter correlation coefficients (_mc, _pcc)	For any set of parameter values, absolute values larger than about 0.95 may indicate that the parameters cannot be uniquely estimated. To explore uniqueness, vary starting parameter values and checking for changes in optimized parameter values.
Ranking of correlation coefficients (_pcc)	Ranked from largest to smallest in the ranges ≥0.95, between 0.90 and 0.95, and between 0.85 and 0.90.
Statistics printed in tables labeled "PARAMETER SUMMARY".[3]	
Parameter values (_pa, _pasub, _paopt, _pc)	When parameter estimation converges, optimized parameter values and the next four items in this table constitute a linear uncertainty analysis of the parameter values. Unreasonable values may indicate problems with observations or the model.
Parameter standard deviations (_pc)	Standard deviations on optimized parameter values indicate the precision with which the values are estimated.
Parameter coefficients of variation (_pc)	Provides a dimensionless measure of the precision with which the parameters are estimated which can be used to compare the precision of parameters with different dimensions.
Parameter 95% linear individual confidence intervals[4] (_pc)	Given normally distributed residuals, reasonable parameter values, a satisfactory model fit, and a linear model, linear confidence intervals reflect the uncertainty of optimal parameter values. Test linearity with the MODEL_LINEARITY code.
Reasonable limits (_pc)	Parameter values and their linear confidence intervals are compared against reasonable ranges supplied by the user.

[1] THIS FONT indicates labels used in the output. _xx is the relevant data-exchange file extension
[2] Printing controlled from the UCODE_Control_Data input block (see Table 14).
[3] If there are log-transformed parameters, parameter values and confidence intervals are printed first transformed as in the regression in a table labeled "PARAMETER SUMMARY", and then as all native values with label "PARAMETER SUMMARY NATIVE". The latter generally is of most interest.
[4] Calculated as usual. For example, see Hill and Tiedeman (in press, eq. 7-8).

Table 31. Using the data-exchange files created by UCODE_2005 that contain data sets for graphical residual analysis of model fit and sensitivity analysis.

[The files are produced when DataExchange=yes in the UCODE_Control_Data input block. See example files presented in Appendix C. x axis, horizontal axis; dss, dimensionless scaled sensitivity; css, composite scaled sensitivity.]

File-name[1]	Intended graph or analysis	Comments[2]
Model Fit		
_os	Observed in relation to simulated values	Ideally, points lie along a line with a slope of 1.0. Uneven spreading along the line may not indicate problems because the values are not weighted.
_ww	Weighted observed in relation to weighted simulated values.	Ideally, points lie along a line with a slope of 1.0. Uneven spreading may indicate problems.
_ws	Weighted residuals in relation to weighted simulated values. Usually, plot weighted simulated values on the x axis.	Ideally, points are evenly distributed above and below the weighted residual zero axis, which indicates random weighted residuals. Uneven spreading along the zero axis may indicate problems.[3]
_r	The residuals listed in this file can be plotted using any independent variable.	Plot residuals on maps, on hydrographs in relation to time, on three-dimensional images of a contaminant plume... Displays model fit, but with unweighted residuals large values may not indicate problems.[3]
_w	Plot the weighted residuals as suggested for the _r file.	As for _r. Non-random plots suggest model bias. Extreme values and groups of negative or positive values suggest problems.[3] Test with a runs test.
_xyztwr	As for _r and _w.	Convenient listing of information together in one file.
_nm	Normal probability graph of the weighted residuals. The probability values are transformed so that they plot on an arithmetic scale.	Ideally, the weighted residuals fall randomly along a straight line. If not, possibilities include: (1) limited number of values or expected fitting by the regression (test using the _rd and _rg files of table 32, (2) problems are indicated.[3]
Sensitivity analysis		
_ss	Plot sum of squared, weighted residuals	Plot for each iteration. Evaluate performance of the regression.
_so	Bar chart of leverage and dss for each observation	Determine whether a few observations dominate the regression.
_sc	Bar chart of css with ParamName on x axis.	Large values = better support by the regression data. Aspects of the system associated with large values perhaps can be represented with more parameters.
_sd	Bar charts of dss for each parameter with the sequential observation number on the x axis.	Usually, a parameter with large css and many large dss is more reliably estimated than a parameter with a large css and one large dss: the error of one important observation is propagated directly to the estimate.
_sl	Use to compare observations with the same units.	Use to compare the importance of different parameters to simulated values.

[1]File names are formed using the second argument on the command line when running UCODE_2005, followed by a period and the extensions listed here.

[2]The phrase "indicate problems" means that the circumstance described indicates that the processes represented by the data may not be adequately modeled or the data may be biased.

[3]For examples, see Hill and Tiedeman (in press) and references cited therein.

Output Files from RESIDUAL_ANALYSIS for Evaluating Model Fit and Identifying Influential Observations

The RESIDUAL_ANALYSIS program produces five files with extensions #rs, _rd, _rg, _rc, and _rb. The fn.#rs file details some intermediate steps of the program and is primarily accessed for the summary of influential observations based on the Cook's D and DFBetas statistics. If the optional fn.rs input file is not used, these summary tables are the only output in the fn.#rs file other than echoed input.

The _rd and _rg files contain sets of generated random numbers. The number of values in each set equals the number of weighted residuals (including values for all observations and prior information). The _rd file contains uncorrelated values; the _rg file contains values correlated to match the correlations produced through the regression. The _rd and _rg files are comprised of lines that contain the generated random number followed by a normal probability plotting position that is adjusted so that it can be plotted on an arithmetic axis (Hill and Tiedeman, in press, Chapter 6.3.6; Hill, 1994); the lines are ordered from the largest to smallest generated value within each of the four sets. On each line the generated random numbers and plotting positions are followed by the "OBSERVATION NAME" and "PLOT SYMBOL" from the associated observation. The values from the _rd and _rg files typically are presented as normal probability graphs along with normal probability graphs produced using the _nm file.

One Cook's D statistic is calculated for each observation and these are contained in the _rc file (table 23, 24, and 32). The Cook's D statistics can be conveniently presented in a bar chart with the sequential observation number on the horizontal axis, or plotted on a map. Large values identify observations that, if omitted, would cause the greatest changes in the estimated parameter values.

DFBetas statistics are calculated for every observation, for every parameter, and are listed in the _rb file (table 23, 24, and 32). Large values identify observations that are influential in the estimation of the parameter. Values for each parameter can be presented in a bar chart or on a map.

Output Files from UCODE_2005 Prediction Mode

UCODE_2005 can be used to generate predictions and prediction sensitivities using optimal parameter values from a previous regression by specifying prediction=yes and sensitivities=yes in the UCODE_Control_Data input block (Chapter 6). Use the same filename prefix, fn; the main output file is fn.#upred. The underscore files listed in table 20 are produced, and include files containing the predictions and their sensitivities at the optimal parameter values used as input to LINEAR_UNCERTAINTY. Summary statistics printed in fn.#upred for user-defined groups of predictions are listed in Table 33.

Chapter 16: Use of Output from UCODE_2005, RESIDUAL_ANALYSIS, MODEL_LINEARITY, and LINEAR_UNCERTAINTY

Table 32. Use of the files created by RESIDUAL_ANALYSIS that contain data sets for graphical residual analysis.
[Summarized from Hill and Tiedeman (in press, Chapters 6.3.6, and 7.5.2).]

File-name[1]	Intended graph or analysis	Comments
_rd	Normal probability graph of random numbers.	Demonstrates the deviation from a straight line caused by a small number of weighted residuals.[2]
_rg	Normal probability graph of correlated random numbers.	Demonstrates the deviation from a straight line caused by a small number of weighted residuals and(or) by an inappropriate conceptual or constructed model.[2]
_rc	Bar chart of the Cook's D statistics with the sequential observation number of the horizontal axis, or maps of the study area with the statistic plotted at the observation location.	Large values identify observations that, if omitted, would result in greater changes to the estimated parameter values than if other observation were omitted.
_rb	Bar charts of DFBetas statistics for each parameter with the sequential observation number of the horizontal axis, or maps of the study area with the statistic plotted at the observation location.	Large values identify observations with the most influence on each parameter estimate.

[1] File names are formed using the second argument on the command line when running UCODE_2005 and RESIDUAL_ANALYSIS, followed by a period and the extensions listed here and in Table 15.
[2] For examples, see Hill and Tiedeman (in press) and references cited therein.

Table 33. Summary statistics for groups of predictions (after Tonkin and others, 2003).
[$pred_i$, prediction number i in the group; ref_i, reference value for prediction number i in the group; n, the number of predictions in the set; Σ, sum for all predictions in the group; %, percent.]

Statistic name	Column Heading for this Statistic in "fn.#upred"	Equation
Summarize predictions		
Maximum predicted value	LARGEST	$Max(pred_i)$
Minimum predicted value	SMALLEST	$Min(pred_i)$
Average predicted value	AVERAGE[1]	$\Sigma (pred_i)/n$
Compare to reference values		
Maximum difference between reference values and predictions	DIFF-MAX	$Max(ref_i\text{-}pred_i)$
Minimum difference between reference values and predictions	DIFF-MIN	$Min(ref_i\text{-}pred_i)$
Mean difference between reference values and predictions	DIFF-AVG[1]	$\Sigma(ref_i\text{-}pred_i)/n$
Compare to reference values as percent		
Maximum percent difference between reference values and predictions	%-DIFF-MAX	$Max[(ref_i\text{-}pred_i)/ref_i] \times 100$
Minimum percent difference between reference values and predictions	%-DIFF-MIN	$Min[(ref_i\text{-}pred_i)/ref_i] \times 100$
Mean percent difference between reference values and predictions	%-DIFF-AVG	$\Sigma[(ref_i\text{-}pred_i)/ref_i]/n \times 100$
Other comparison to reference values		
Number of predictions less than their reference value	# < REF	
Number of predictions greater than their reference value	# > REF	

[1] Meaningful only if all predictions involved have the same units.

Output Files from LINEAR_UNCERTAINTY for Predictions

LINEAR_UNCERTAINTY prints predictions and 95-percent linear confidence and prediction intervals on the predictions. The sequence of runs needed is described in Chapter 13. The LINEAR_UNCERTAINTY output files are named fn.#linunc and fn._linp, where filename prefix fn is replaced by a label defined on the command line. Sections of the _linp output file are labeled, indicating the type of confidence or prediction interval that follows in tabular format (as described in Table 26). Six labels are used. The first three are for confidence intervals and are:

(1) INDIVIDUAL 95% CONFIDENCE INTERVALS

(2) SIMULTANEOUS 95% CONFIDENCE INTERVALS,

and

(3) UNDEFINED NUMBER OF SIMULTANEOUS 95% CONFIDENCE INTERVALS

The first label is followed by individual confidence intervals.

For the second label, the subsequent interval limits depend on the number of intervals calculated. The intervals get wider as more intervals are calculated until the number of intervals exceeds the number of parameters involved in the calculation. At that point the intervals do not increase in size regardless of how many intervals are calculated and the intervals listed under the second and third heading are the same. Under the second heading, the intervals are calculated using either the Bonferroni or Scheffé d=k equations (Hill and Tiedeman, in press, section 8.4.2); both produce conservative (large) intervals so the smallest of the two is most accurate and therefore listed. Bonferroni equations are used when they are equal.

The third label is followed by Scheffé d=NP confidence intervals.

Labels 4, 5, and 6 are for prediction intervals. The labels are the same as labels 1, 2, and 3 except that the word "confidence" is replaced the word "prediction".

The theory for calculating confidence and prediction intervals is discussed by Hill (1994) and Hill and Tiedeman (in press, Chapter 8). The linearity assumption of these confidence and prediction intervals needs to be evaluated using MODEL_LINEARITY.

Output Files from MODEL_LINEARITY for Testing Linearity

MODEL_LINEARITY calculates the modified Beale's measure of model linearity (Cooley and Naff, 1990; Hill, 1994, p. 47; Hill and Tiedeman, in press) and statistics that indicate the magnitude of the nonlinearity of each parameter. When regression is performed and DataExchange=yes, UCODE_2005 produces an _b1 output file, which is then used by UCODE_2005 in a separate run with linearity=yes to produce an _b2 file. Generally, the user does not access either of these files. MODEL_LINEARITY uses these files to produce the MODEL_LINEARITY output file fn.#modlin, an example of which is distributed electronically with UCODE_2005. Information related to interpretation of the output is included at the end of the file. Hill (1994) or Hill and Tiedeman (in press) explain the modified Beale's measure and the information printed in the MODEL_LINEARITY output file.

Chapter 16: Use of Output from UCODE_2005, RESIDUAL_ANALYSIS, MODEL_LINEARITY, and LINEAR_UNCERTAINTY

Chapter 17: NONLINEAR CONFIDENCE INTERVALS AND ADVANCED EVALUATION OF RESIDUALS AND NONLINEARITY

Christensen and Cooley (2005) present a set of capabilities for MODFLOW-2000 that include advanced methods for evaluating residuals, nonlinearity, and uncertainty based largely on Cooley (2004). Advanced evaluation of uncertainty uses nonlinear confidence intervals. Christensen and Cooley (2005) call this set of capabilities the UNC Process of MODFLOW-2000. These capabilities are accessed in the MODFLOW-2000 framework through three post-processing codes executed externally to MODFLOW-2000. These codes are called RESAN2-2K, CORFAC-2K, and BEALE2-2K.

The methods of Christensen and Cooley (2005) have been adapted for use with UCODE_2005 through the addition of two modes of UCODE_2005 and the development of three codes. The two modes are advanced-test-model-linearity mode and nonlinear-uncertainty mode. The three codes are modified from the three programs listed above, and are called, respectively, RESIDUAL_ANALYSIS_ADV, CORFAC_PLUS, and MODEL_LINEARITY_ADV.

In this chapter, we describe how to obtain results using the new capabilities. For further information about the methods, including the equations being solved, readers are referred to Christensen and Cooley (2005). In addition, nonlinear intervals without correction factors are described by Hill and Tiedeman (in press). As experience with these methods evolves, their utility and limitations will become better understood and likely will be the topic of a variety of publications. We include them here in UCODE_2005 so that this experience can evolve more quickly.

A fundamental assumption of the methods documented by Christensen and Cooley (2005) is that the weighting on the prior information correctly reflects its true error. In general, this means that the methods are not applicable to underdetermined parameterizations in which the weighting on the prior is assigned to produce a tractable regression problem and is inconsistent with an analysis of errors in the prior information.

Two capabilities described by Christensen and Cooley (2005) are not included here:

(1) The approximate procedure governed by IMAN (as described on their p. 40).

(2) The ability to have correction factors calculated with different values specified for input variables WG and VP (as described on their p. 59). Here, WG=VP is used.

Omission (1) means that nonlinear confidence intervals for some poorly posed problems that could be solved using the UNC Process of MODLOW-2000 may not be able to be solved using UCODE_2005. It was omitted from this version of UCODE_2005 due to time constraints. Omission (2) is not expected to affect many users.

What Christensen and Cooley (2005) refer to as Scheffé intervals are called Scheffé d=NP intervals in this document.

This chapter first describes the project flow in terms of the sequence of runs required. It then lists the new data-exchange files in a table for quick reference. The remainder of the chapter is spent documenting the three new codes – RESIDUAL_ANALYSIS_ADV, CORFAC_PLUS, and MODEL_LINEARITY_ADV – and the two additional modes of UCODE_2005.

Project Flow Using the Advanced Capabilities

Project flow using the advanced capabilities is best understood by considering the files produced and used by the UCODE_2005 modes and the computer programs involved. Table 34 and figure 17 provide this information. The new data-exchange files listed in table 34 are described briefly in table 35.

As discussed for the LINEAR_UNCERTAINTY post-processing code in Chapter 15, the parameters defined for the UCODE_2005 nonlinear-uncertainty mode listed in table 34 may differ from the parameters defined for the calibration. To use the modified parameter definitions, the parameter definitions need to be changed and the first run listed in table 34 needs to be the sensitivity-analysis mode of UCODE_2005. See steps 1 and 2 for execution of LINEAR_UNCERTAINTY in Chapter 15 for additional information.

Another circumstance that requires an initial sensitivity-analysis run is when parameters have become inactive during the regression. Parameters can become inactive through dynamic omission of insensitive parameters activated by the OmitInsensitive keyword of the Reg_GN_Controls input block or constraints activated using the Constrain keyword in the Parameter_Groups or Parameter_Data input block. If parameters have become inactive, the user needs to replace this parameter-estimation mode run with a sensitivity-analysis mode run. In the Parameter_Data input block, for the inactivated parameters two choices are possible: (a) set Adjustable=no to omit these parameters from further consideration or (b) set Adjustable=yes to include them. In either circumstance, assign these parameters values from the end of the parameter-estimation run or assign them what are thought to be more reasonable values.

As far as the authors know, no one has attempted to calculate nonlinear intervals using parameters that were not estimated in the regression. Considering that each limit of each interval is obtained by regression, it is likely that including such parameters will cause convergence problems. Yet if these parameters are important to predictions of interest, their inclusion may be important to obtaining meaningful prediction uncertainty. It is easy to include the effect of such parameters using the linear intervals produced using the code LINEAR_UNCERTAINTY, as described in Chapter 15. An example is shown in Appendix C.

The results of the program RESIDUAL_ANALYSIS_ADV are not used by other programs (table 34), but are used to evaluate model fit and, thus, the validity of the results produced by the other programs.

Chapter 17: Advanced Evaluation of Residuals, Nonlinearity, and Uncertainty
-- Project Flow Using the Advanced Capabilities--

The second sensitivity-analysis mode run listed in table 34 is only needed if the _init and _init._** files were created using optimized parameter values. This occurs under the following circumstances:

(a) The first run listed in table 34 is a UCODE_2005 parameter-estimation mode run, and starting parameter values are close to final parameter values.

(b) The first run listed in this table is a UCODE_2005 sensitivity-analysis run.

If the sensitivity-analysis run to create init files is needed, the following are required:

(i)   In the UCODE_Control_Data input file, set keywords Sensitivity=yes, Optimize=no, and CreateInitFiles=yes.

(ii) Use the Parameter_Data or Parameter_Values input block to define a set values for the estimated parameters that are different than the estimated values, but not so different that the sum of squared, weighted residuals exceeds the sum-of-squared, weighted residuals value of the nonlinear intervals. The latter are printed in the _int* data-exchange files, where the * is replaced by "conf" for confidence intervals and "pred" for prediction intervals. Generally a set of values from the _pasub file created by the UCODE_2005 parameter-estimation mode can be used. Add values for added parameters.

Some of the UCODE_2005 runs shown in table 34 need process model results for calibration conditions, some need process model results for prediction conditions, and some need both. Comments for setting up runs for both are presented in the section on the advanced-test-model-linearity mode in this chapter.

Table 34. Project flow for the advanced analyses and nonlinear confidence intervals documented in Chapter 17, as illustrated through input and output files.
[Shading identifies runs other than UCODE_2005. *, replaced by 'conf' for confidence intervals or 'pred' for prediction intervals. In _init._** the ** represents dm, mv, su, and, when prior information is defined, supri. (calibration), UCODE_2005 mode needs calibration conditions; (prediction), the mode needs prediction conditions; (calibration and prediction), the mode needs calibration and prediction conditions.]

UCODE_2005 Mode or CODE NAME [1]	Input files [2,3]	Output files [2,4]
UCODE_2005 Parameter-Estimation or Sensitivity-Analysis[5] Mode (calibration)	--	#uout _b1, _dm, _init, _init._**, _mv, _os, _paopt, [_pr], _su, [_supri], _w, _ws, _wt, [_wtpri], _ww
RESIDUAL_ANALYSIS	rs (optional) _dm, _mv, _su, [_supri], _w, _wt, [_wtpri]	#resan _rb, _rc, _rd, _rg
UCODE_2005 Sensitivity-Analysis Mode with CreateInitFiles=yes[6] (calibration)	--	#ucreateinitfiles _init, _init._**
RESIDUAL_ANALYSIS_ADV	rsadv (optional) _init, _init._**, _paopt, _ws, _wt, _wtpri, _ww	#resanadv _rdadv [7]
UCODE_2005 Test-Model-Linearity Mode (calibration)	--	_b2
MODEL_LINEARITY	_b1, _b2	#modlin
UCODE_2005 Prediction Mode[8] (prediction)	_paopt	#upred _p, _pv, _spu
LINEAR_UNCERTAINTY	_dm, _mv, _p, _pv, _spu	#linunc _linp
CORFAC_PLUS	corfac _mv, [_p, _spu], _su, [_supri], _wt, [_wtpri]	#corfac_* _b1adv*, _b3*, _cf*, _cfsu
UCODE_2005 Advanced-Test-Model-Linearity Mode (calibration and prediction)	_b1adv*, _b3*, _cfsu	#umodlinadv_* _b2adv*, _b4*
MODEL_LINEARITY_ADV	_b1adv*, _b2adv*, _b3, _b4*, _cf*, _cfsu, _dm, _init, _os, _paopt, [_pr], _su, [_supri], _wt, [_wtpri]	#modlinadv_*
UCODE_2005 Nonlinear-Uncertainty Mode (calibration and prediction)[9]	_init , _init._**, _cf*, _cfsu, _paopt, _pv	#unonlinint_* _int*, _int*par, _intwr

[1] UCODE_2005 modes are described in table 3. Other capitalized names identify other computer codes documented in this report.

[2] The file-name extensions are listed. Main output file extensions begin with #; main input file extensions are composed only of letters; data-exchange file extensions begin with _. The file-name prefix is fn., where fn is defined on the command line. fn needs to be the same for the entire sequence of runs listed in this table. Bracketed file extensions, such as [_pr], identify data-exchange files that may not be needed or produced; the files involve prior information (_pr, _supri, _wtpri, _init._supri) or predictions (_p, _spu).

[3] For UCODE_2005 mode input files, the data-exchange files included are those produced by a previous model run listed in this table.

[4] The output files listed are the main output files and, for UCODE_2005 modes, data-exchange files used by one or more subsequently listed modes or programs. For the other runs, all output files are listed.

[5] This sensitivity-analysis mode run needs to have the optimized parameter values listed in the UCODE_2005 main input file. In the UCODE_Control_Data input block keyword CreateInitFiles=no is needed by default or designation. See table 3 for other sensitivity-analysis mode requirements.

[6] CreateInitFiles is a keyword in the UCODE_Control_Data input block. Often this run is not needed. See the text for an explanation of this run.

[7] _rdadv is used to plot ordered weighted residuals and their confidence intervals. An example is shown in Appendix C.

[8] This run is only needed if there are predictions. If intervals are only to be calculated on parameters, this can be skipped.

[9] Prediction conditions need not be simulated if intervals are to be calculated only on parameter values.

A

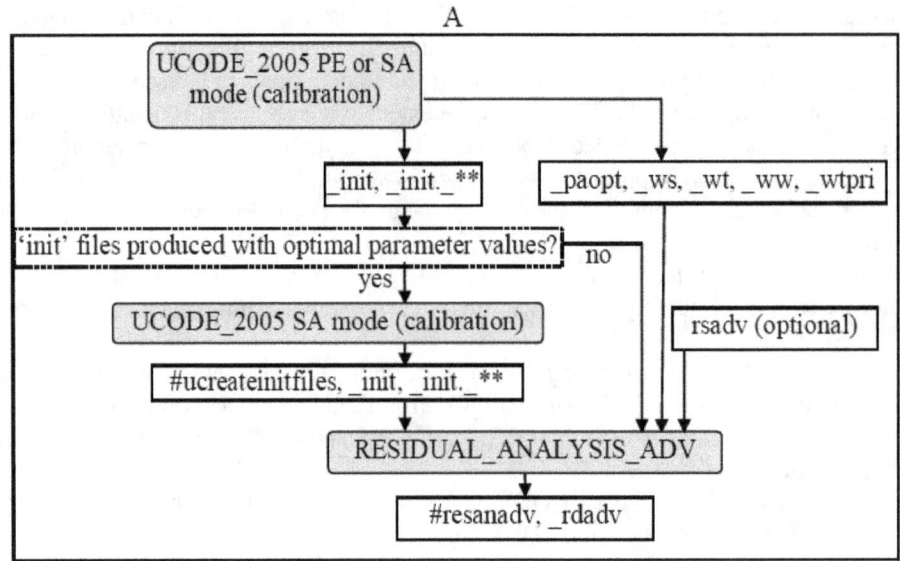

B

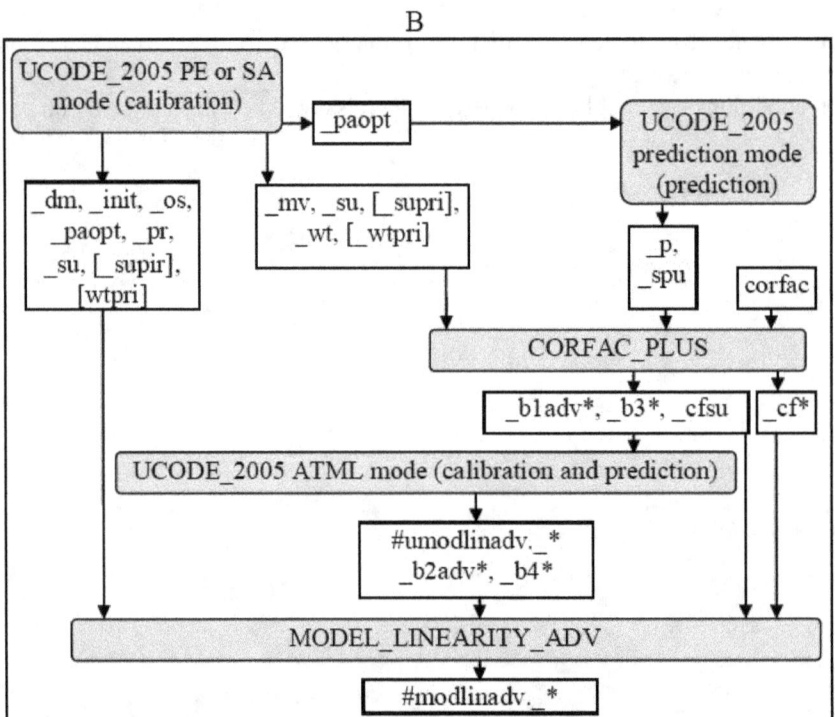

Figure 17. Flowcharts with UCODE_2005 and CORFAC_PLUS runs for
(A) RESIDUAL_ANALYSIS_ADV, (B) MODEL_LINEARITY_ADV, and
(C) nonlinear intervals. Shaded boxes identify code executions. Only output files
needed by subsequent codes are listed. UCODE_2005 modes are indicated as PE,
parameter-estimation; SA sensitivity-analysis; ATML, advanced-test-model-
linearity; NU, nonlinear-uncertainty. (calibration) indicates that calibration
conditions need to be simulated; (prediction) indicates that prediction conditions
need to be simulated; (calibration and prediction) indicates both are needed.

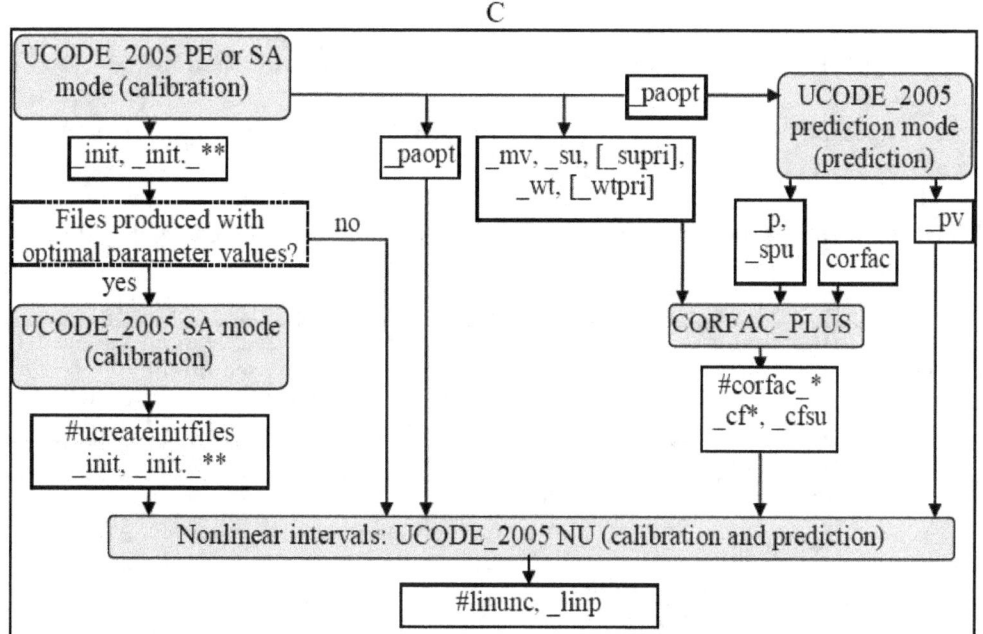

Figure 17 -- continued

Data-Exchange Files for Advanced Capabilities

The new capabilities described in this chapter produce the data-exchange files listed in table 35.

Table 35. Contents of data-exchange files produced and used by the UCODE_2005 and the codes discussed in Chapter 17.

[* is replaced by 'conf' for confidence intervals, 'pred' for prediction intervals.]

File Extension	Description	`Column Tags`. Surrounded by double quotes in header. Most `Column Tags` are used literally. `ParamName` and `ObsName` are replaced by user defined names. Numbers are added to emphasize the order.
The following is produced by RESIDUAL_ANALYSIS_ADV		
_rdadv	RESIDUAL_ ANALYSIS_ ADV results	Eight column tags[1] Number of data rows = Number of observations
The following are produced by CORFAC_PLUS		
_b1adv[2]	Parameter sets for _b2adv.	`ParamName1 ParamName2 ParamName3 ...` Number of data rows = 2 × number of estimated parameters
_b3*[2]	Parameter sets for _b4*	`ParamName1 ParamName2 ParamName3 ...` Number of data rows = number of intervals
_cf*	Correction factors	Six column tags.[3] Number of data rows = Number of intervals
_cfsu[2]	Sensitivities	`Item Name Pred=1,Par=2  ParamName1 ParamName2 ...` Sensitivities for each quantity for which intervals are calculated.
The following are produced by the UCODE_2005 Advanced-Test-Model-Linearity Mode		
_b2adv*[2]	Values simulated using _b1adv.	Number of columns = number of observations + 1 Number of data rows = 2 × number of estimated parameters
_b4*[2]	Values simulated using _b3	`ObsName1  ObsName2  ... PREDICTION`[4] Number of data rows = Number of intervals
The following are produced by the UCODE_2005 Nonlinear-Uncertainty Mode		
_int*	Nonlinear interval limits	Eight column tags[5] Number of data rows = one for each interval limit
_int*wr	Weighted residuals at interval limits	[6]Number of column tags = 2 + one for each interval limit Number of data rows = Number of observations + Number of Prior Information

_int*par[2]	Parameter values for the limits in _int*	`INTERVAL LIMIT ParamName1 ParamName2 ...`
		 `NAME      FLAG`
		 Number of data rows = one for each interval limit

[1] The _rdadv column tags are (1) ORDERED WEIGHTED RESIDUALS, (2) ORDERED SIMULATED WEIGHTED RESIDUALS (OSWR), (3) STD DEV OF OSRW, (4) 2*(STD DEV), (5) CUMULATIVE

PROBABILITY, (6) PROBABILITY PLOTTING POSITION, (7) OBSERVATION or PRIOR NAME, (8) PLOT-SYMBOL. See additional comments in the section "Output Files for RESIDUAL_ANALYSIS_ADV".

[2] Up to 500 values are printed as a set using long lines in the output file. Additional values are printed in subsequent additional sets in the same file. Each set has the headers shown and the observation name and plot symbol in the first two columns.

[3] The _cf* file columns tags are (1) INTERVAL NAME, (2) Pred=1 Par=2, (3) PRED VALUE, (4) MEAS VAR, (5) CF INDIVID, (6) CF SIMULT.

[4] For _b4* the column headers are the observation names followed by the header "PREDICTION".

[5] The _int* file columns tags are (1) INTERVAL NAME, (2) PLOT SYMBOL, (3) LIMIT IDENTIFIER, (4) CONFIDENCE LIMIT, (5) SUM OF SQUARED RESIDUALS, (5) OBJECTIVE-FUNCTION GOAL, (6) PERCENT DEVIATION FROM GOAL, (7) INDIVIDUALorSIMULTANEOUS, (8) ITERATIONS. Under "LIMIT IDENTIFIER", negative and positive integers identify lower and upper limits, respectively. The absolute value is the same for limits calculated for each interval. The absolute values increase incrementally for each interval such that for the first interval the values are -1 and 1, for the second they are -2 and 2, and so on.

[6] The _int*wr files columns tags are (1) OBSERVATION or PRIOR INFORMATION NAME (2) PLOT SYMBOL. These are followed by one or two columns for each interval, depending on how many limits are calculated for each interval. If there are two (WhichLimits=Both in the Reg_GN_NonLinInt input block), the first is labeled −Name and the second is labeled +Name. "Name" is replaced by an interval name. If the first character is −, the column is comprised of lower limits; if the first character is +, the column is comprised of upper limits. If WhichLimits=Lower, only columns of lower limits are printed and the first character of each of the labels is −. If WhichLimits=Upper, only columns of upper limits are printed and the first character of each of the labels is +.

RESIDUAL_ANALYSIS_ADV: Advanced Residual Analysis

The RESIDUAL_ANALYSIS_ADV program can be executed after successfully running UCODE_2005 in parameter-estimation or sensitivity-analysis mode (table 34).

RESIDUAL_ANALYSIS_ADV results can be used to conduct an analysis that addresses the issues of concern addressed by the _rd and _rg files produced by the RESIDUAL_ANALYSIS program (Chapter 15) and the _nm file produced by UCODE-2005 in some modes (see tables 3 and 15). However, the analysis provided by RESIDUAL_ANALYSIS_ADV differs in how problems are identified – weighted residuals that fall outside calculated intervals indicate significant lack of model fit. An example of the type of graph that can be produced using the results of RESIDUAL_ANALYSIS_ADV is shown by Christensen and Cooley (2005, p. 44).

RESIDUAL_ANALYSIS_ADV also calculates the intrinsic model nonlinearity measure.

RESIDUAL_ANALYSIS_ADV does not completely replace RESIDUAL_ANALYSIS, and in general it is advantageous to run both programs. Specifically, the DFBETAS and Cook's D statistics provided in the _rb and _rc files produced by RESIDIAL_ANALYSIS remain useful. In addition, the _ws, _ww, and _r files produced by UCODE_2005 remain useful.

Execution

The RESIDUAL_ANALYSIS_ADV run command is of the form:

path:\ RESIDUAL_ANALYSIS_ADV.exe fn

where:

path:\ = the relative or absolute path to the RESIDUAL_ANALYSIS_ADV.exe on your computer (alternatively you could specify this in your system path variable)

fn = filename prefix for data-exchange files that were generated by the regression and prediction executions of UCODE_2005 and the execution of CORFAC_PLUS (spaces are not allowed in fn, even on operating systems that allow spaces in filenames)

Most input files for RESIDUAL_ANALYSIS_ADV are produced by running UCODE_2005 in parameter-estimation or sensitivity mode, as shown in table 34 and figure 17A. In addition, some circumstances require the second UCODE_2005 sensitivity-analysis mode that creates init files.

RESIDUAL_ANALYSIS_ADV produces files fn.#resanadv and fn._rdadv.

User-Prepared Input File (Optional)

Input for RESIDUAL_ANALYSIS_ADV includes the previously generated data-exchange files listed in table 34. It also includes an optional user-created file with filename fn.rsadv, where fn is the filename prefix specified on the command line. This file includes up to four input blocks described in the following sections:

> Options
> RESIDUAL_ANALYSIS_ADV_Control_Data
> Mean_True_Error
> Matrix_Files

All of the input blocks are optional.

Options Input Block (Optional)

The Options input block controls the information written to the main RESIDUAL_ANALYSIS_ADV output file. It has a single keyword:

Verbose   - Flag that controls what is written to the RESIDUAL_ANALYSIS_ADV main output file as follows. The default is Verbose=3 to provide information for new applications and users, but Verbose=0 is suggested for most circumstances.

Verbose	Output
0	No extraneous output.
1	Warnings.
2	Warnings, notes.
3 (default)	Warnings, notes, echo selected input.
4	Warnings, notes, echo all input. Includes all values read from model output files.
5	Warnings, notes, echo all input, plus some miscellaneous information. Includes all values read from model output files.

Example of an Options Input Block for the RESIDUAL_ANALYSIS_ADV program:

```
BEGIN Options Keywords
Verbose=0
END Options
```

RESIDUAL_ANALYSIS_ADV_Control_Data Input Block (Optional)

The RESIDUAL_ANALYSIS_ADV_Control_Data input block provides control of several features of the RESIDUAL_ANALYSIS_ADV program. The need for the last two input blocks of the fn.rsadv depend on keywords defined in this input block. The keywords of the RESIDUAL_ANALYSIS_ADV_Control_Data input block are listed in table 36.

Table 36. Keywords of the RESIDUAL_ANALYSIS_ADV_Control_Data input block. [Input block format is defined in Chapter 5.]

Keyword	Description	Default	Options
NSETS	Number of sets of random numbers to be generated	1000	A positive integer.[1]
SEED	Seed used to generate random numbers	104857	A positive integer.[2]
TVAR	Theoretical error variance. When not equal to zero, this value is used instead of the calculated error variance.[3]	0.0	A positive real number.
READ_ET	If READ_ET=yes, read the Mean_True_Error input block.[4] If READ_ET=no, the values are set to 0.0.	No	Yes/No
READ_COV	If READ_COV=yes, read one matrix from the MATRIX_FILES block.[5]	No	Yes/No
The following keywords control printing to the fn.#resanadv output file. Specify Yes to print the data described; No not to print the data. These keywords do not affect production of the data-exchange file _rdadv.			
PRINT_IRMATRIX	Print the (I-R) matrix.[6]	No	No/Yes
PRINT_ SIMWGTRESIDUALS	Print all sets of the generated weighted residuals	No	No/Yes
PRINT_PAR_ VAR_COV_MATRIX	Print the variance-covariance matrix of the estimated parameters	No	No/Yes
PRINT_SQRT_WT	Print the square root of the weight matrix as a second matrix.	No	No/Yes
PRINT_ UNSCALED_SENS	Print unscaled sensitivities for the observations	No	No/Yes

[1] Commonly in the 100's or 1000's.

[2] Needs to be between 1 and 1,048,575.

[3] For more information, see Christensen and Cooley (2005, p. 46). Used to calculate NSETS realizations of random numbers with the theoretical distribution of the weighted residuals. These are then used to calculate intervals with which weighted residuals are compared.

[4] The ET values of Christensen and Cooley (2005, p. 47), which are defined by Cooley (2004, p. 21, eq. 3-31) and represent the errors produced because the model is a simplified representation of the system in that small-scale variability is not represented. These values tend toward zero as more small-scale variability is represented. If READ_ET=no, ET values for all observations are set to 0.0

[5] If the statistics specified in the observation input blocks are not based on an analysis of the observation errors, statistics that reflect the observation errors need to be specified here in the form of a variance-covariance matrix. That is, if the statistics specified in the observation input blocks do not result in weighting that is proportional to the inverse of the variance-covariance matrix of the true errors, then variance-covariance matrix needs to be specified in the MATRIX_FILES input block. (Christensen and Cooley, 2005, p. 15)

[6] This matrix times a multiplicative constant equals the variance-covariance matrix of the weighted residuals of the observations (Christensen and Cooley, 2005, p. 15)

Example of a RESIDUAL_ANALYSIS_ADV_Control_Data input block:

```
#For keywords not listed, default values are assigned.
BEGIN RESIDUAL ANALYSIS ADV CONTROL DATA KEYWORDS
NSETS = 1000
SEED  = 10059
STDDEV = 0.01000000
END RESIDUAL ANALYSIS ADV CONTROL DATA
```

Mean_True_Error Input Block (Optional)

Mean true errors can be specified for any of the observations and reflect the errors produced because the model does not represent small-scale variations in the actual system. They are not provided for prior information because it is assumed that the weighting on the prior correctly reflects the true error in the prior information. See the comment at the beginning of this chapter.

Mean true errors can be determined using Monte Carlo simulations with small grid spacing that allows explicit representation of the small-scale variations. Computational constraints may require these simulations to represent only a portion of the full system being simulated.

The Mean_True_Error input block has two keywords.

MTEName   - ObsName from the Observation_Data or the Derived_Observations input blocks used in the regression. Default=0.0.

MTEValue   - The ET values of Christensen and Cooley (2005, p. 47), which are defined by Cooley (2004, p. 21, eq. 3-31) and represent the errors produced because the model does not represent small-scale variations present in the actual system. Default=0.0.

These values tend toward zero as the model represents more small-scale features. The values are set to zero if the RESIDUAL_ANALYSIS_ADV_Control_Data input block specifies READ_ET=no.

Example of a Mean_True_Error input block:

```
BEGIN MEAN TRUE ERROR TABLE
NROW=5  NCOL=2  COLUMNLABELS
mtename    mtevalue
F1    0.66151
F2    0.57163
F3    0.57791
F4    0.65444
F5    0.70896
END MEAN TRUE ERROR
```

Matrix_Files Input Block (Optional)

If the weighting of observations is not defined as being proportional to the variance-covariance matrix of the true errors, the variance-covariance matrix of the true errors needs to be defined here. This is communicated to the program using the keyword READ_COV=yes in the RESIDUAL_ANALYSIS_ADV_Control_Data.

The Matrix_Files input block is as described in Chapter 10. Here some brief instructions and some restrictions on this use of the Matrix_Files capability are provided.

There are two keywords in the Matrix_Files input block.

MatrixFile   - Name or path of the file from which one matrix is read. (Up to 2,000 characters; case sensitivity depends on the operating system).

NMatrices   - Number of matrices to be read from MatrixFile. For the RESIDUAL_ANALYSIS_ADV_Control_Data input block, NMatrices=1 is needed.

Example of a Matrix_Files input block:

```
BEGIN MATRIX FILES
matrixfile=Vmatrix   NMATRICES=1
END MATRIX FILES
```

Output Files for RESIDUAL_ANALYSIS_ADV

RESIDUAL_ANALYSIS_ADV produces two output files, as listed in table 34: the main output file and the data-exchange file with file extensions #resanadv and _rsadv, respectively.

The main output file contains quantities described by Christensen and Cooley (2005, p. 43-45) that can be used to test for intrinsic nonlinearity (mean weighted residual, slope of weighted residual, and an intrinsic nonlinearity measure) and to test weighted residuals for indications of non-normality (correlation and probability of correlation). The output printed for the test case described in Appendix C without the porosity parameter defined is shown below

```
.
.   lines of output file not listed
.
MEAN WEIGHTED RESIDUAL (EM) --------- = 0.99833E-01 (SHOULD ~ 0.00)
SLOPE (SLP) ------------------------- = 6.1693E-05  (SHOULD ~ 0.00)
(SLOPE OF THE PLOT OF WEIGHTED RESIDUALS
VS WEIGHTED SIMULATED EQUIVALENTS)
INTRINSIC NONLINEARITY MEASURE (QINT) = 0.63002 (SHOULD BE << 23.910)
.
.   lines of output file not listed
```

```
.
CORRELATION (CED) --------------- = 0.98549
PROBABILITY OF CORRELATION (PROB) = 0.44100
99% CONFIDENCE LIMIT (CL99) ----- = 0.99563
95% CONFIDENCE LIMIT (CL95) ----- = 0.99389
90% CONFIDENCE LIMIT (CL90) ----- = 0.99304
```

The mean weighted residual (EM) and slope (SLP) should be close to zero; no critical values are defined by Cooley and Naff (2005). The intrinsic nonlinearity measure should be much less than the sum-of-squared weighted residuals calculated for the optimal parameter values, which is listed in parentheses following the statistic. This statistic is used to indicate whether the intrinsic model nonlinearity is large enough to make the correction factors calculated by CORFAC_PLUS inaccurate.

The correlation (CED) is between the weighted residuals and the means of the synthetic residuals and is described by Christensen and Cooley (2005, p. 17) as a generalization of the R_N^2 statistic described by Hill and Tiedeman (in press) and Hill (1998). R_N^2 is printed in the output file of MODEL_LINEARITY. The correlation printed by RESIDUAL_ANALYSIS_ADV should be close to 1.00.

The probability of correlation (PROB) is the probability that a correlation is equal to or smaller than CED given that the weighted residuals are normally distributed as described by Christensen and Cooley (2005, eq. 30 or 31). The smaller the value of PROB, the greater the indication that the weighted residuals are affected by model bias and intrinsic nonlinearity. The confidence limits listed are the values that the correlation would need to exceed to achieve a probability of the stated amount or higher. The 95-percent confidence limit for the correlation equals CL95. In the example, the correlation would need to be larger than or equal to 0.99389 to achieve a probability of 0.95 or higher that the weighted residuals are consistent with the theoretical distribution.

The correlation is the correlation between the weighted residuals and the weighted simulated values, obtained from the UCODE_2005 produced data-exchange files _ws and _ww, respectively. Ideally the correlation is close to 1.00. The intrinsic nonlinearity measure, which is different that the intrinsic model nonlinearity produced by MODEL_LINEARITY_ADV,

The data-exchange file with file extension _rsadv contains eight columns of numbers. The labels printed on the first line of the file are listed in table 35, and listed below with additional comments as needed.

(1) ORDERED WEIGHTED RESIDUALS,
(2) ORDERED SIMULATED WEIGHTED RESIDUALS (OSWR), the mean of many realizations of random numbers correlated as expected based on the fitting process of the regression,
(3) STD DEV OF OSRW, the standard deviation of the realizations for each observation,
(4) 2*(STD DEV), two times the standard deviations of the realizations,
(5) CUMULATIVE PROBABILITY, for plotting on a normal probability axis,
(6) PROBABILITY PLOTTING POSITION, for plotting on an arithmetic axis,

(7) OBSERVATION or PRIOR NAME, name of the observation or prior information, and
(8) PLOT-SYMBOL, an integer that can be used to control the symbol used for plotting.

This file can be used to create the plots similar to that shown in figure 1 of Christensen and Cooley (2005, p. 44). An example is shown in Appendix C.

CORFAC_PLUS: Correction Factors and Data for Analysis of Linearity

Program CORFAC_PLUS can be used to calculate the correction factors of Cooley (2004) and Christensen and Cooley (2005), and prepares files used by the UCODE_2005 advanced-test-model-linearity mode, the program MODEL_LINEARITY_ADVANCED, and the UCODE_2005 nonlinear-uncertainty mode. Thus, it accomplishes the calculations performed by CORFAC-2K of Christensen and Cooley (2005), and produces some files for analysis of model linearity. This functionality is reflected in the name CORFAC_PLUS. Table 34 shows how CORFAC_PLUS fits into a sequence of runs.

It is common to calculate nonlinear intervals on predictions without correction factors. For example, such intervals are calculated by PEST's Predictive Analyzer capability (Doherty, 2004). To obtain such results, the input blocks needed by CORFAC_PLUS are the Correction_Factor_Data input block (where the only keyword that needs to be included is IntervalType), and the Prediction_List or input block to define the predictions for which nonlinear confidence intervals are to be calculated.

Correction factors are intended to quantify the effects on measures of uncertainty such as confidence or prediction intervals of (1) intrinsic nonlinearity and (2) small-scale variability of system characteristics not represented in the model. As explained by Cooley (2004) and Christensen and Cooley (2005), it is assumed that the model correctly represents the spatial and temporal average of all system characteristics. Neglecting the small-scale variability adds uncertainty to parameters and predictions. The correction factors account for at least part of this added uncertainty.

The types of intervals and equations used to calculate the correction factors calculated by CORFAC_PLUS are shown in table 37.

The correction factors are derived assuming that the intrinsic nonlinearity and combined intrinsic nonlinearity are both negligible (Cooley, 2004; Christensen and Cooley, 2005, p. 55). The correction factors are used by MODEL_LINEARITY_ADV to check that these assumptions are adhered to, and, if they are not, to indicate the severity of the departure from these assumptions.

Table 37. Equations from Christensen and Cooley (2005) used to calculate correction
factors for different types of intervals.

Type of interval	Is the variance-covariance matrix of the true errors available?[1]		
	Yes: Used in regression	Yes: Read in Matrix_Files input block	No
Individual Confidence Interval	[2]1.0	Eq. 68	Eq. 83
Individual Prediction Interval	Eq. 70	Eq. 70	Eq. 84
Scheffé d=NP Simultaneous Confidence Interval	Eq. 66	Eq. 66	Eq. 82
Scheffé d=NP Simultaneous Prediction Interval	Not available[3]	Not available[3]	Not available[3]

[1] It can be available either because (i) the weighting used in the regression accounts for errors in observations, so that the weight matrix equals or is proportional to the inverse of the variance-covariance matrix of the true errors, or (ii) the variance-covariance of the true errors is read in the Matrix_Files input block.

[2] The correction factor equals 1.0.

[3] No theory has been developed for simultaneous prediction intervals.

Execution

The CORFAC_PLUS run command is of the form:

path:\ CORFAC_PLUS.exe   fn

where:

path:\ = the relative or absolute path to the CORFAC_PLUS.exe on your computer
(alternatively you could specify this in your system path variable)

fn = filename prefix for data-exchange files that were generated by the regression and
prediction executions of UCODE_2005 (spaces are not allowed in
fn, even on operating systems that allow spaces in filenames).

Input for CORFAC_PLUS includes previously generated files listed in table 34 and
figure 17B and C, and a user created file fn.corfac. CORFAC_PLUS produces a main
output file with file extension #corfac and the data-exchange files listed in table 34.

User-Prepared Input File (Required)

CORFAC_PLUS user-prepared input file needs to be named fn.corfac, where fn is defined on the command line. This file includes up to five input blocks, some of which are optional. The input blocks are described below and need to appear in the following order:

Input block name	Comment
Options	Optional
Correction_Factor_Data	Required to define IntervalType
Prediction_List Parameter_List	At least one of these input blocks is needed
Matrix_Files	Optional

Options Input Block (Optional)

The Options input block controls the information written to the main CORFAC_PLUS output file. It has a single keyword:

Verbose   - Flag that controls what is written to the CORFAC_PLUS main output file as follows. The default is Verbose=3 to provide information for new applications and users, but Verbose=0 is suggested for most circumstances.

Verbose	Output
0	No extraneous output.
1	Warnings.
2	Warnings, notes.
3 (default)	Warnings, notes, echo selected input.
4	Warnings, notes, echo all input. Includes all values read from the process-model output files.
5	Warnings, notes, echo all input, plus some miscellaneous information. Includes all values read from the process-model output files.

Example of an Options Input Block for the CORFAC_PLUS program:

```
BEGIN Options Keywords
Verbose=5
END Options
```

Correction_Factor_Data Input Block (Required)

The Correction_Factor_Data input block is composed of five keywords.

ConfidenceOrPrediction - confidence: confidence intervals are calculated. prediction: prediction intervals are calculated. Confidence and prediction intervals are defined in Chapter 3. In a single run of CORFAC_PLUS, correction factors are calculated for one type of interval. No default

RegressionUsedTrueCov – yes: In the preceding UCODE_2005 parameter-estimation or sensitivity-analysis mode run (the first run listed in table 34), the statistics specified in the observation input blocks are consistent with an analysis of the observation errors (see table 37). no: The statistics are not representative of the observation errors. Default=no.

Read_Cov - yes: Read a matrix from the Matrix_Files input block. Default=no.

If the statistics specified for the observations are not based on an analysis of the observation errors, statistics can be specified here in the form of a variance-covariance matrix using the Matrx_Files input block. That is, if the statistics specified in the observation input blocks do not result in weighting that is intended to be proportional to the inverse of the variance-covariance matrix of the true errors, that variance-covariance matrix needs to be specified in the Matrix_Files input block. (Christensen and Cooley, 2005, p. 15). If RegressionUsedTrueCov=yes and Read_Cov=yes, an error message is returned and execution is stopped.

EffectiveCorrelation - The upper limit of the spatial correlation. Default=0.8.

The value of EffectiveCorrelation only is used if Read_Cov=no and RegressionUsedTrueCov=no. The spatial correlation is defined as **C** by Cooley (2004, p. 48-49) and Christensen and Cooley (2005, p. 26-28). When needed, the upper limit of the spatial correlation (EffectiveCorrelation) is used to calculate the variable 'a' in equation 76 of Christensen and Cooley (2005, p. 27) and to approximate correction factors (Christensen and Cooley, 2005, p. 26-28).

Read_ObsPredCov - yes: Read a matrix of covariances between observations and predictions (**C** of Christensen and Cooley, 2005, p. 12, eq. 18) using the Matrix_Files input block. no: Do not read a matrix; the second moments are set to 0.0. Default=no.

Example of a Correction_Factor_Data input block:

```
BEGIN CORRECTION_FACTOR_DATA KEYWORDS
  IntervalType = CONFIDENCE
  RegressionUsedTrueCov = yes
END CORRECTION FACTOR DATA
```

Prediction_List Input Block (this block, the next block, or both are needed)

The Prediction_List input block lists the predictions for which correction factors are calculated by CORFAC_PLUS and for which nonlinear intervals can be calculated in a subsequent execution of the nonlinear-uncertainty mode of UCODE_2005. Either the Prediction_List input block or the Parameter_List input block described in the next section, or both, need to be provided.

Predictions listed in the Prediction_List input block need to also be listed in the preceding prediction mode run of UCODE_2005. However, not all of the predictions need to be listed here -- only the ones for which correction factors are to be calculated. The prediction names listed here are checked against the prediction names listed for the prediction mode; the comparison is case-insensitive.

The Prediction_List input block has three keywords.

PredName — Prediction name for which a correction factor is to be calculated. Each name needs to be one of the names listed in the Prediction_Data input block of the associated UCODE_2005 run.

The following two keywords define the statistic needed to calculate prediction intervals. If ConfidenceOrPrediction=Confidence in the Correction_Factor_Data input block, the following two keywords are not used to calculate correction factors.

MeasStatistic — A statistic used to calculate the variance with which the predicted quantity could be measured. Used by the nonlinear-uncertainty mode of UCODE_2005 to calculate prediction intervals. Default = variance printed in the _pv file. (see below)

MeasStatFlag — Character string that defines how the corresponding MeasStatistic is used to calculate the measurement error for the prediction. Default=VAR. Options are:

MeasStatFlag	Variance is calculated as
VAR	MeasStatistic
SD	$(\text{MeasStatistic})^2$

If Blockformat KEYWORDS is selected by designation or default, keywords related to a prediction need to be grouped together and follow the related PredictionName. The PredictionName keyword needs to be the first keyword on a new line. PredictionName and associated keywords are repeated to list multiple predictions.

For ConfidenceOrPrediction=Prediction in the Correction_Factor_Data input block, by default the MeasStatistic for the predictions listed are obtained from the _pv file as variances (this file contains variances calculated using the values of MeasStatistic specified in the Prediction_Data input block from the Prediction mode). Values for MeasStatistic and MeasStatFlag specified in this Prediction_List input block replace values from the _pv file. Only the predictions for which variances are to be changed need to have values assigned for MeasStatistic and MeasStatFlag.

To obtain results for a combination of confidence and prediction intervals, set ConfidenceOrPrediction=Prediction and set MeasStatistic=0.0 for the predictions for which confidence intervals are desired.

Example of a Prediction List input block:

```
BEGIN PREDICTION LIST TABLE
nrow=6 ncol=2 COLUMNLABELS MEASSTATFLAG=VAR
PREDICTIONNAME MeasStatistic
G1              10.907
G2              7.5765
G3              7.2347
G4              5.2386
G5              3.4021
G6              6.6283
END PREDICTION LIST
```

Parameter_List Input Block (this block, the last block, or both are needed)

The Parameter_List input block lists the parameters for which correction factors are calculated by CORFAC_PLUS and for which nonlinear intervals are calculated in a subsequent execution of the nonlinear-uncertainty mode of UCODE_2005. Either the Parameter_List input block or the Prediction_List input block described in the previous section, or both, must be provided.

Parameters listed in the Parameter_List input block need to have had sensitivities calculated in the preceding prediction mode run of UCODE_2005. Not all of the parameters need to be listed here -- only the ones for which correction factors and nonlinear intervals are to be calculated. The list of parameters is checked using ParamNames; the comparison is case-insensitive.

The Parameter_List input block has three keywords.

ParameterName   - Parameter name for which a correction factor, and possibly a nonlinear interval, is to be calculated. Each name needs to be one of the names listed in the Parameter_Data input block of the associated UCODE_2005 run.

The following two keywords define the statistic needed to calculate prediction intervals. If ConfidenceOrPrediction=Confidence in the Correction_Factor_Data input block, the following two keywords are not used to calculate correction factors.

Prediction intervals are rarely calculated for parameters. If they are, MeasStatistic and MeasStatFlag define the expected precision with which unbiased parameter values can be measured.

MeasStatistic   - A statistic used to calculate the variance with which the parameter could be measured. Used by the nonlinear-uncertainty mode of UCODE_2005 to calculate prediction intervals. Rarely used for parameters. Default=0.0.

MeasStatFlag   - Character string that defines how the corresponding MeasStatistic is used to calculate the variance of the error with which the parameter value can be measured. No default. Options are:

MeasStatFlag	Variance is calculated as
VAR	MeasStatistic
SD	$(MeasStatistic)^2$

If Blockformat KEYWORDS is selected by designation or default, keywords related to a parameter need to be grouped together and follow the related ParameterName. The ParameterName keyword needs to be the first keyword on a new line. ParameterName and associated keywords are repeated to list multiple predictions.

Example of a Parameter_List input block:

```
BEGIN PARAMETER LIST TABLE
nrow=2 ncol=1 COLUMNLABELS
PARAMETERNAME
HK_1
HK_2
END PARAMETER LIST
```

Matrix_Files Input Block (Optional)

Up to two matrices are read as indicated by keywords Read_Cov and Read_ObsPredC in the Correction_Factor_Data input block. If both of the keywords are set to 'yes', two matrices are read and the matrix associated with Read_Cov needs to be first.

This Matrix_Files input block performs as described in Chapter 10. Here some brief instructions and some restrictions on this use of the Matrix_Files capability are provided.

There are two keywords in the Matrix_Files input block.

MatrixFile   - Name or path of the file from which one or more matrices are read. (Up to 2,000 characters; case sensitivity depends on the operating system).

NMatrices   - Number of matrices to be read from MatrixFile.

Example of a Matrix_Files input block:

```
BEGIN MATRIX FILES
matrixfile=V-VPmatrix  NMATRICES=2
END MATRIX FILES
```

Output Files for CORFAC_PLUS

Each run of CORFAC_PLUS produces five output files, as listed in table 34: the main output file with file extension #corfac_* and data-exchange files with file extensions _b1adv*, _b3*, _cf*, and _cfsu. The * is replaced by 'conf' or 'pred' depending on the designations for keyword ConfidenceOrPrediction in the Correction_Factor_Data input block.

The * is replaced by 'conf' when ConfidenceOrPrediction=Confidence and 'pred' when ConfidenceOrPredicion=Prediction.

The CORFAC_PLUS main input file, with file extension #corfac_*, is very similar to the output file from the CORFAC_2K program described by Christensen and Cooley (2005, p. 56-57, 117-121). Readers are referred to that source for further information.

The _b1adv* data-exchange file contains sets of parameter values, where the number of sets equals two times the number of estimated parameters.

The_b3* data-exchange file also contains sets of parameters values. The number of sets equals the number of items listed in the Prediction_List and Parameter_List input blocks in the preceding run of CORFAC_PLUS.

The _cf* data-exchange file contains correction factors.

The _cfsu data-exchange file contains sensitivities for the quantities for which intervals are calculated evaluated at the optimal parameter values.

The Advanced-Test-Model-Linearity Mode of UCODE_2005

The advanced-test-model-linearity mode of UCODE_2005 calculates simulated values for each of the sets of parameter values listed in the _b1adv* and _b3* data-exchange files, where * is replaced by 'conf' for confidence intervals and 'pred' for prediction intervals. It produces the files _b2adv* and _b4*. See tables 34 and 35 and figure 17B.

In most respects, the UCODE_2005 main input file needs to be identical to a forward mode run. Exceptions are as follows.

1. Set keyword LinearityAdv=conf or pred in the UCODE_Control_Data input block (table 3; Chapter 6).

2. The advanced-test-model-linearity mode requires information for both observations and predictions. Items 3 to 5 describe how to accomplish this.

3. Change the Model_Command_Lines input block of Chapter 6 as needed to run the process models required to obtain simulated equivalents to the observations and predictions. Generally this means using the methods discussed for the Model_Command_Lines input block to run the process model as in the UCODE_2005 parameter-estimation mode and the prediction mode runs.

4. Use the input blocks described in Chapter 8 to define both observations and predictions. When starting with a UCODE_2005 main input file from a parameter-estimation mode run, this means adding input blocks for predictions. Generally these can simply be copied from the preceding prediction mode run.

5. Change the input blocks described in Chapter 11 to interact with the process model input and output files as needed to obtain results for both calibration and prediction conditions. When starting with a parameter-estimation mode run main input file, this usually means integrating the lines from the prediction mode run into these input blocks. In the Model_Output_Files input block, each ModOutFile keyword needs to be associated with a Category keyword, with Category=Obs for observations or Category=Preds for predictions. Any output files that include both observations and predictions need to be listed twice: once with the instruction file to read the values required for observations with Category=Obs, and once with the instruction file to read the values for predictions with Category=Preds.

Input and output files for the advanced-test-model-linearity mode are listed in table 34.

MODEL_LINEARITY_ADV: Advanced Evaluation of Model Linearity

MODEL_LINEARITY_ADV differs substantially from MODEL_LINEARITY, which is documented in Chapter 15. MODEL_LINEARITY_ADV calculates total model nonlinearity, intrinsic model nonlinearity, and combined intrinsic model nonlinearity. The types of nonlinearity measures and a few comments are provided in table 38.

The role of model predictions highlighted in table 38 is important if the predicted quantity and(or) prediction conditions are substantially different than the quantities used to calibrate the model and (or) calibration conditions. In ground-water modelling, this occurs when a model calibrated with heads and flows is used to simulate transport, or when pumping conditions change considerably. If there are substantial differences, linearity measures that do not account for predictions may underestimate the effects of nonlinearity on calculated intervals. For more information, see Christensen and Cooley (2005), Cooley (2004), and Hill and Tiedeman (in press).

Execution

The MODEL_LINEARITY_ADV run command is of the form:

path:\ MODEL_LINEARITY_ADV.exe fn

where:

path:\ = the relative or absolute path to the RESIDUAL_ANALYSIS_ADV.exe on your computer (alternatively you could specify this in your system path variable)

fn = filename prefix for data-exchange files that were generated by the regression and prediction executions of UCODE_2005 and the execution of CORFAC_PLUS (spaces are not allowed in fn, even on operating systems that allow spaces in filenames)

Input Files for MODEL_LINEARITY_ADV

MODEL_LINEARITY_ADV relies on previously generated data-exchange files listed in Table 34, as shown in Figure 17B. Execution needs to be preceded by four runs:

1. UCODE_2005 parameter-estimation mode or sensitivity mode using optimal parameter values.
2. UCODE_2005 prediction mode.
3. CORFAC_PLUS.
4. UCODE_2005 advanced-test-model-linearity mode.

No user-generated input files are needed or possible.

213

Table 38. The model nonlinearity measures listed in the MODEL_LINEARITY_ADV main output file.

Name[1]	Comment	Critical Values
Total model nonlinearity[2] (p. 18-19)	Nonlinearity of the simulated equivalents of the observations with respect to the parameters. Equals the sum of intrinsic model nonlinearity and parameter effects nonlinearity. Parameter effects nonlinearity is defined as nonlinearity that could be eliminated by a parameter transformation, though the transformation is unknown.	>1.0 highly nonlinear 0.09 to 1.0 nonlinear 0.01 to 0.09 moderately nonlinear <0.01 effectively linear
Intrinsic model nonlinearity[3] (p. 19-20)	Nonlinearity of the simulated equivalents of the observations that cannot be eliminated by parameter transformation.	
Combined intrinsic model nonlinearity[3] (eq. 48 or 56)	One value is produced for each interval. Nonlinearity of the simulated equivalents of the observations and the prediction that cannot be eliminated by the same parameter transformation.	≤0.01
Combined intrinsic nonlinearity – maximum sum (p. 24, 48)	The largest of BMI+2BMF0 or |BMI-2BMG0|. BMI=combined intrinsic model nonlinearity, BMF0=what BMI would equal if the simulated equivalents to observations were linear, and BMG0=what BMI would equal if the predictions were linear. Measures the magnitude of intrinsic model nonlinearity and combined intrinsic model nonlinearity. If it is small, linear and nonlinear confidence or prediction intervals are similar.	<0.09 (conservative limit; larger values may be okay)

[1] Equation or page numbers listed are where the statistic is defined in Christensen and Cooley (2005). For combined intrinsic model nonlinearity, eq. 48 and 56 apply for confidence and prediction intervals, respectively.

[2] Conclusions about nonlinearity are similar to those reached using the modified Beale's measure calculated by MODEL_LINEARITY.

[3] Called model intrinsic nonlinearity and model combined intrinsic nonlinearity, respectively, by Cooley (2004, p. 36, 56) and Christensen and Cooley (2005).

Output File for MODEL_LINEARITY_ADV

MODEL_LINEARITY_ADV produces: fn.#modlinadv, which lists the nonlinearity measures listed in Table 38 and associated critical values. Total model nonlinearity and intrinsic model linearity are printed about halfway down the file and appear as follows.

```
##############################################################
##########
##########    TOTAL NONLINEARITY (BNT)......... = 45.433
##########    INTRINSIC NONLINEARITY (BNI)...... = 0.12908
##########
##########CRITICAL VALUES FOR BOTH MEASURES:
##########    >1.0 highly nonlinear
##########    0.09 to 1.0 non-linear
##########    0.01 to 0.09 moderately nonlinear
##########    <0.01 effectively linear
##############################################################
```

Combined intrinsic model nonlinearity and associated statistics are printed at the bottom of the file and appear as follows.

```
COMBINED INTRINSIC NONLINEARITY
    Standard linear intervals are good approximations
          for predictions with values <= 0.01
   INTERVAL NO.         BMI          INTERVAL NO.        BMI
1  AD10_X        4.4280          4  A100_X       4.0473
2  AD10_Y        489.68          5  A100_Y       53.153
3  AD10_Z        9.9946          6  A100_Z       13.687

   COMBINED INTRINSIC NONLINEARITY AS IF F WERE LINEAR
   INTERVAL NO.         BMF0         INTERVAL NO.       BMF0
1  AD10_X        4.4275          4  A100_X       4.0467
2  AD10_Y        489.68          5  A100_Y       53.152
3  AD10_Z        9.9940          6  A100_Z       13.686

   COMBINED INTRINSIC NONLINEARITY AS IF G WERE LINEAR
   INTERVAL NO.         BMG0         INTERVAL NO.       BMG0
1  AD10_X        0.54578E-03     4  A100_X       0.61157E-03
2  AD10_Y        0.45722E-03     5  A100_Y       0.67193E-03
3  AD10_Z        0.61090E-03     6  A100_Z       0.10208E-02

       COMBINED INTRINSIC NONLINEARITY - MAX. SUM
Correction factors are not affected by combined intrinsic model
  linearity if this value <0.09. This limit is conservative;
        larger values may not affect correction.
   INTERVAL NO.         BMIMAX       INTERVAL NO.       BMIMAX
1  AD10_X        13.283          4  A100_X       12.141
2  AD10_Y        1469.0          5  A100_Y       159.46
3  AD10_Z        29.983          6  A100_Z       41.059
```

The Nonlinear-Uncertainty Mode of UCODE_2005

Nonlinear intervals are calculated using the nonlinear-uncertainty mode of UCODE_2005 (table 3). Nonlinear intervals can not be calculated using a post-processor such as LINEAR_UNCERTAINTY because each limit of each interval requires that a full regression be completed.

The regression procedure conducted by the nonlinear-uncertainty mode seeks to find parameter values that produce the extreme values of each quantity for which intervals are being calculated. For example, for a prediction, the goal is to find parameter values that produce the largest and smallest values of the prediction. The parameter values also need to result in simulated equivalents of the observations that produce a specified objective-function value. The specified value is called the objective-function goal in this report.

The objective-function goal is determined based three factors: (1) whether the interval is a confidence or prediction interval, (2) whether the interval is an individual or simultaneous interval, and (3) whether correction factors are all set to 1.0 or determined by CORFAC_PLUS.

The quantities for which intervals are calculated are defined in the CORFAC_PLUS input file, using the Prediction_List and Parameter_List input blocks.

An execution of the nonlinear-uncertainty mode of UCODE_2005 has either ConfidenceOrPrediction=Confidence or ConfidenceOrPrediction=Prediction in the Red_GN_NonLinInt input block described below in this section. Files from a CORFAC_PLUS run with the same option specified in the Correction_Factor_Data input block need to be present in the directory in which the UCODE_2005 nonlinear-uncertainty mode is run, using the same filename prefix on the command line.

Preparatory Steps

The steps by which the user prepares for nonlinear interval calculation are as follows. It is generally most useful to modify a copy of the related UCODE_2005 parameter-estimation mode main input file. If intervals are calculated on predictions (the Prediction_List input block is used in the CORFAC_PLUS input file), parts of a related UCODE_2005 prediction mode main input file are needed for the nonlinear-uncertainty mode main input file.

1. For the UCODE_2005 nonlinear-uncertainty mode to calculate intervals on predictions, the nonlinear-uncertainty mode needs to follow successful completion of at least three, and possibly four, other runs. Other runs may be important to model analysis, but are not needed to calculate nonlinear confidence intervals. The three required runs are:

(a) UCODE_2005 parameter-estimation (the first run in table 34),

(b) UCODE_2005 prediction mode, and

(c) CORFAC_PLUS.

In addition, the second sensitivity-analysis mode run listed in table 34 is needed if the _init and _init._** files generated by (a) are inadequate. They can be inadequate in three situations: (1) the parameter-estimation mode starting parameter values equal the optimal values, (2) the parameter-estimation mode run is a replaced by a sensitivity-analysis mode run with the parameter values equal to the optimal parameter values (this is needed if the parameter definition is changed, as discussed in the section "Project Flow Using the Advanced Capabilities" of this chapter), or (3) the parameter-estimation mode starting values produce a very large value of the objective function. Problems are indicated if the _rdadv file contains intervals that are so wide that the weighted residuals plot on a vertical line because the scale on the horizontal axis is so large. If problems are suspected, run the second sensitivity-analysis mode listed in table 34 to produce init files. Use parameter values from an intermediate parameter-estimation iteration. These parameter values are listed in the data-exchange file with extension _pasub produced by the parameter-estimation mode run. The data-exchange file from the same run with extension _ss can be used to determine the objective-function values produced by each set of parameter values in the _pasub file. For additional information, see the section at the beginning of Chapter 17 entitled "Project Flow Using the Advanced Capabilities".

If nonlinear intervals are calculated only on parameters (the Parameter_List input block is used in the CORFAC_PLUS input file and the Prediction_List input block is not), the situation is the same as that described above except that the prediction run (b) is not needed.

2. In the UCODE_Control_Data input block, include keyword NonLinearIntervals=yes. Make sure the other keywords that control the UCODE_2005 mode are set to "no" by default or designation. Specifics are provided in table 3 and Chapter 6.

3. If intervals are to be calculated on predictions, the UCODE_2005 nonlinear-uncertainty mode main input file needs to be able to run calibration and prediction conditions. See instructions in the section on the advanced-test-model-linearity mode of UCODE-2005 earlier in this chapter. In addition, if derivatives-interface files described in Chapter 13 are used for the observations and the predictions, the following steps are needed.

(a) If separate runs of the process model are used to produce sensitivity-equation sensitivities, the files containing the sensitivities need to be merged into one file. In the Options input block of Chapter 6, use the PathToMergedFile keyword to define the file. In the Merge_Files input block of Chapter 6, list the files to be merged. Modify the derivatives-interface file described in Chapter 13 to read the merged file; this generally requires adding the prediction names to the list of dependent variables and changing NDEP accordingly (see table 13).

217

4. As noted in the description of the Parameter_Values input block in Chapter 7, better regression convergence can sometimes be achieved using different starting values. If AlternateStartValues=yes in the Reg_GN_NonLinInt input block, parameter values listed in the Parameter_Values input block are used as the starting values. These values may be different for different runs of the nonlinear-uncertainty mode. The starting parameter values need to be within the parameter confidence region, so can not be too far from the optimal values. If the AlternateStartValues=no, parameter values from the _paopt data-exchange file are used.

5. In the Parameter_Data input block, the values assigned for ScalePval are used in the nonlinear-uncertainty mode as described for keyword TolIntP of the Ref_GN_NonLinInt input block described next. Typically, the absolute values of the estimated parameter values listed in the _paopt data-exchange file are used.

6. If the defaults are not adequate, add a Reg_GN_NonLinInt input block to the UCODE_2005 main input file. This input block is described in the following section of this report.

7. The observation and prior information input blocks from the parameter-estimation or sensitivity-analysis mode run (the first run listed in table 34) are required without modification.

8. List the predictions for which nonlinear intervals are to be calculated in the prediction input blocks described in Chapter 8. The prediction input blocks from the prediction mode run (table 34) can be modified to include only predictions for which intervals are to be calculated. If new predictions are of interest, they need to be added to the prediction mode input file and the prediction mode and subsequent runs needs to be repeated before proceeding (see step 1).

9. Execute UCODE_2005. A regression is performed for each interval limit, so execution times can be long.

Multiple computers can be used to reduce the time required to obtain nonlinear confidence intervals. There are two possible ways of using multiple computers:

(a) For each interval limit, when sensitivities are calculated by perturbation, each process-model run can be sent to a different computer using the parallel capabilities described in Chapter 12.

(b) The calculations for each interval limit are completely independent of one another. Thus they can be calculated using different computers. This is not supported by the parallelization capabilities included in UCODE_2005; instead, it can be accomplished simply by running UCODE_2005 on different computers. The only difference between the runs would be the predictions defined in the input blocks defined in Chapter 8 (see item 8 above) and possibly the designation of the WhichLimits keyword in the Reg_GN_NonLinInt input block described in the following section.

Reg_GN_NonLinInt Input Block (Optional)

The Reg_GN_NonlinInt input block can be added to the UCODE_2005 main input file to define options for calculating nonlinear intervals. It follows the Reg_GN_Controls input block and precedes the Model_Command_Lines input block. Keywords only need to be included to use something other than the defaults, and the input block can be omitted if all defaults are used.

Data from the Reg_GN_Controls input block used to calculate nonlinear interval limits include the designations for keywords MaxIter, MaxChange, MaxChangeRealm, and MqrtDirection. The value of MaxChange is often important to obtaining convergence for interval limits. In the event of convergence problems try decreasing MaxChange and increasing MaxIter.

The Parameter_Data input block allows MaxChange values to be specified, but these values are ignored when calculating nonlinear interval limits. Only the MaxChange value in Reg_GN_Controls is used to calculate nonlinear interval limits.

The quasi-Newton updating, dynamic omission of parameters, and trust region features for which keywords are defined in the Reg_GN_Controls input block are not available for calculating nonlinear intervals. Designations for these keywords are not used.

ConfidenceOrPrediction – confidence: confidence intervals are calculated. prediction: Prediction intervals are calculated. Default=confidence.

Confidence and prediction intervals are defined in Chapter 3. In a single run of the nonlinear-uncertainty mode, only one type of interval is calculated. If ConfidenceOrPrediction=Confidence, and data-exchange file _cfconf CORFAC_PLUS does not exist, correction factors are not used to calculate the confidence intervals. Similarly, if ConfidenceOrPrediction=Prediction, and data-exchange file _cfpred does not exist, correction factors are not used to calculate the prediction intervals. The _cfconf and _cfpred data-exchange files can be produced by CORFAC_PLUS.

IndividualOrSimultaneous – individual: individual intervals are calculated. simultaneous: simultaneous intervals are calculated. Default=individual.

Individual and simultaneous intervals are defined in Chapter 3. In a single run of the nonlinear-uncertainty mode, only one type of interval is calculated.

Simultaneous intervals are all Scheffé d=NP intervals, which are accurate when the number of intervals exceeds the number of parameters, and tend to be too large for fewer intervals.

For prediction intervals (ConfidenceOrPrediction=prediction), only individual intervals can be calculated; if IndividualOrSimultaneous=simultaneous, the designation is ignored.

WhichLimits   - The interval limits to be calculated. Options are Lower, Upper or Both. Default=Both.

TolIntP   - Tolerance based on parameter values: a limit of a confidence interval is achieved if the maximum absolute value of the fractional change in parameter values between iterations is less than $\text{ScalePval}_j \times \text{TolIntP}$. ScalePval is a keyword defined for the Parameter_Data input block and the values specified there are used in this calculation. The subscript j refers to the jth parameter. Default=0.001.

TolIntS   - Tolerance based on model fit. Convergence is achieved when the fractional change over two iterations is less than TolIntS, calculated as $|S(\theta_{r+1}) - S(\theta_r)| / S(\theta_r) + |S(\theta_r) - S(\theta_{r-1})| / S(\theta_{r-1}) < \text{TolIntS}$. Christensen and Cooley (2005, p. 41) found that a value of one-tenth of TolIntP worked well in the problems they tested. While the relation to TolIntP depends on parameter sensitivity, the default is based on their experience. Default=$0.1 \times \text{TolIntP}$.

TolIntY   - Tolerance based on the change in the value of the computed interval limit. Convergence is achieved when the value of the limit has not changed more than TolIntY times the average of the last two values of the limit. That is, $|2 \times (g(\gamma\theta_{r+1}) - g(\gamma\theta_r)) / (g(\gamma\theta_{r+1}) + g(\gamma\theta_r))| < \text{TolIntY}$, where $g(\gamma\theta)$ is the iteratively calculated value at the limit, r is the iteration counter, and $\gamma\theta_r$ is a vector of parameter values at iteration r. Default=0.001.

CorrectionFactors   - yes: read and use correction factors calculated by CORFAC_PLUS. no: do not read or use correction factors (equivalent to setting the correction factors to 1.0). Default=no.

Even when correction factors are not used, CORFAC_PLUS needs to be run before the UCODE_2005 nonlinear-uncertainty mode run.

AlternativeStartValues - yes: use the parameter values listed in the Parameter_Values input block to start the regression for all interval limits. no: use the estimated parameter values listed in file fn._paopt. Use AlternativeStartValues=yes to investigate uniqueness of the interval limits and possibly to aid convergence. Default=no.

Calculating a Subset of the Interval Limits

When calculating a set of interval limits, two circumstances may result in the need to recalculate some of the limits. First, some limits may not converge. This is apparent because there is a message in the main output file and a message is written instead of a value in the _int* file. Second, convergence may be obtained but the large discrepancy between the sum-of-squared weighted residuals and the desired value is too large. Satisfactory limits often can be achieved by changing the MaxStep in the Reg_GN_Controls input block or changing the convergence criteria in the Ref_GN_NonLinInt input block.

Several runs may be required to obtain some limits, and it is computationally intensive to recalculate all the limits for each of these runs. There are two ways to restrict the calculated limits. First, the WhichLimits keyword in the Reg_GN_NonLinInt input block can be used to calculate only upper or lower limits.

Second, some of the quantities for which limits are calculated can be omitted. To omit predictions, the predictions need to be removed from (1) the input blocks in the main UCODE_2005 input file, (2) the instruction file listed in the Model_Output_Files input block, and (3) the Prediction_List input block in the CORFAC_PLUS input file. In some circumstances it may be convenient to change the process model so that the predictions are not calculated. This primarily occurs when the Standard File option is being used to read from model output files, and it is easier to change the predictions produced by the process model than to construct a more complicated instruction file. For example, if MODFLOW-2000 is used and predictions are read from an _os file and sensitivities are read from the _su file using the standard file option, one option is to change the MODFLOW run so that these files list results for the predictions of cocern.

To omit parameters, the parameters need to be omitted from the Parameter_List input block of the CORFAC_PLUS input file and keyword NonLinearInterval defined for the Parameter_Data input block of the UCODE_2005 main input file needs to be changed.

Nonlinear-Uncertainty Mode Output Files

Results are reported in the output files fn.#unonlinint_*, _int*, fn._intwr, and _int*par, as listed in table 34. The symbol * is replaced by 'conf' when confidence intervals are calculated and 'pred' when prediction intervals are calculated.

The main output file is composed of an initial echo of input information followed by information from each iteration of the nonlinear regression process used to calculate each limit of each interval. The main output file can be large. Generally modelers need to look at the output file only for debugging and to explore the iterations of limits for which nonlinear regression did not converge.

Files with filename extensions _intconf, _intconfpar, and _intconfwr are produced if confidence intervals are calculated and files with extensions _intpred, _intpredpar, and _intpredwr are produced if predictions intervals are calculated.

The confidence intervals in the _intconf file can be individual or simultaneous intervals, depending on the designation of IndividualOrSimultaneous in the Reg_GN_NonLinInt Input Block. The prediction intervals in the _intpred file are always individual intervals regardless of how keyword IndividualOrSimultaneous is defined.

Chapter 17: Advanced Evaluation of Residuals, Nonlinearity, and Uncertainty
--The Nonlinear Uncertainty Mode of UCODE_2005--

The parameter values for each limit are listed in the _int*par file. The weighted residuals calculated using the parameter values for each limit are listed in the _int*wr file, where * is replaced by 'conf' or 'pred'. These files can be used to check calculations for the nonlinear interval limits, and the weighted residuals can be used to check for intrinsic nonlinearity as described by Christensen and Cooley (2005, p. 11, 36).

It is important to determine whether all of the intervals are associated with the appropriate value of the sum-of-squared weighted residuals. This can be evaluated by comparing the values in the fifth column of the _intconf or _intpred data-exchange files against the critical value listed in the sixth column. The percent difference is listed in the seventh column.

For more information about the data-exchange files, see table 35.

Chapter 18: REFERENCES

Anderman, E.R. and Hill, M.C., 1999, A new multistage groundwater transport inverse method: Presentation, evaluation, and implications: Water Resources Research., v. 35, No. 4, p. 1053-1063.

Barlebo, H.C., Hill, M.C., Rosbjerg, D., and Jensen, K.H., 1998, Concentration data and dimensionality in groundwater models: Evaluation using inverse modeling: Nordic Hydrology, v. 29, p. 149-178.

Barth, G.R. and Hill, M.C., 2005. Numerical methods for improving sensitivity analysis and parameter estimation of virus transport simulated using sorptive-reactive processes: Journal of Contaminant Hydrology, v. 76, p. 251-277.

Barth, G.R. and Hill, M.C., in press, Parameter and observation importance in modeling virus transport in saturated systems – Investigations in a homogenous system: Journal of Contaminant Hydrology.

Belsley, D.A, Kuh, E., and Welsch, R.E, 1980, Regression diagnostics, Identifying influential data and source of collinearity: John Wiley & Sons, New York, 292 p.

Christensen, Steen and Cooley, R.L, 1999, Simultaneous confidence intervals for a steady-state leaky aquifer groundwater flow model: Advances in Water Resources Special Section on Model Calibration and Reliability Evaluation, v. 22, No. 8, p. 807-817.

Christensen and Cooley, 2005, User guide to the UNC process and three utility programs for computation of nonlinear confidence and prediction intervals using MODFLOW-2000: U.S. Geological Survey Techniques and Methods Report 2004-1349, 186p.

Christensen, Steen, Rasmussen, K.R., and Moeller, K., 1998, Prediction of regional ground-water flow to streams: Ground Water, v. 36, No. 2, p. 351-360.

Cook, R.D. and Weisberg, S., 1982, Residuals and influence in regression: Chapman and Hall, New York, 230 p.

Cooley, R.L., 2004, A theory for modeling ground-water flow in heterogeneous media: U.S. Geological Survey Professional Paper 1679, 220 p.

Cooley, R.L. and Hill, M.C., 1992, A comparison of three Newton-like nonlinear least-squares methods for estimating parameters of ground-water flow models, in Russell, T.F., Ewing, R.E., Brebbia, C.A., Gray, W.G., and Pinder, G.F., eds., Computational Methods in Water Resources 9th, vol. 1: Numerical methods in water resources, Elsevier, p. 379-386.

Cooley, R.L. and Naff, R.L., 1990, Regression modeling of ground-water flow: U. S. Geological Survey Techniques in Water-Resources Investigations, book 3, Chapter B4, 232 p.

D'Agnese, F.A., Faunt, C.C., Turner, A.K, and Hill, M.C., 1997, Hydrogeologic evaluation and numerical simulation of the Death Valley Regional ground-water flow system, Nevada and California: U.S. Geological Survey Water-Resources Investigations Report 96-4300, 124 p.

D'Agnese, F.A., Faunt, C.C. Hill, M.C., and Turner, A.K., 1999, Death Valley regional ground-water flow model calibration using optimal parameter estimation methods and geoscientific information systems: Invited paper for a Special Section of Advances in Water Resources on Model Calibration and Reliability Evaluation For Ground-Water Systems, A. Leijnse and M.C. Hill, eds., v. 22, No. 8, p. 777-790.

Dennis, J.E. and Schnabel, R.B., 1996, Numerical methods for unconstrained optimization and nonlinear equations: Society for Industrial and Applied Mathematics, Philadelphia, USA, 378 p.

Doherty, John, 2004, PEST-2000: Corinda, Australia, Watermark Computing, http://www.sspa.com/PEST/index.html

Draper, N.R. and Smith, H., 1998, Applied Regression Analysis (3rd ed.): John Wiley & Sons, New York, 706 p.

Eberts, S.M. and George, L.L., 2000, Regional ground-water flow and geochemistry in the midwestern basins and arches aquifer system in parts of Indiana, Ohio, Michigan, and Illinois: U.S. Geological Survey Professional Paper 1323-C, 103 p.

Gailey, R.M., Gorelick, S.M., and Crowe, A.S., 1991, Coupled process parameter estimation and prediction uncertainty using hydraulic head and concentration data: Advances in Water Research, v. 14, No. 5, p. 301-314.

Harbaugh, A.W., Banta, E.R., Hill, M.C., and McDonald, M.G., 2000, MODFLOW-2000, the U.S. Geological Survey modular ground-water model, User's guide to the modularization concepts and the ground-water flow process: U.S. Geological Survey Open-File Report 00-92, 121 p. http://water.usgs.gov/nrp/gwsoftware/modflow2000/modflow2000.html

Helsel, Dennis, 2004, Nondetects and data analysis, Statistics for censored environmental data: Wiley and Sons, New York, 250p.

Hill, M.C., 1990, Preconditioned conjugate gradient 2(PCG2), a computer program for solving ground-water flow equations: U.S. Geological Survey Water-Resources Investigations Report 90-4048, 43 p.

Hill, M.C., 1992, A computer program (MODFLOWP) for estimating parameters of a transient, three-dimensional, ground-water flow model using nonlinear regression: U.S. Geological Survey Open-File Report 91-484, 358 p.

Hill, M.C., 1994, Five computer programs for testing weighted residuals and calculating linear confidence and prediction intervals on results from the ground-water parameter estimation computer program MODFLOWP: U.S. Geological Survey Open-File Report 93-481, 81 p.

Hill, M.C., 1998, Methods and guidelines for effective model calibration: U.S. Geological Survey Water-Resources Investigations Report 98-4005, 90 p.

Hill, M.C., Banta, E.R., Harbaugh, A.W., and Anderman, E.R., 2000, MODFLOW-2000, the U.S. Geological Survey modular ground-water model, User's guide to the Observation, Sensitivity, and Parameter-Estimation Process and three post-processing programs: U.S. Geological Survey Open-File Report 00-184, 209 p. http://water.usgs.gov/nrp/gwsoftware/modflow2000/modflow2000.html

Hill, M.C. and Tiedeman, C.R., in press, Effective groundwater model calibration, with analysis of sensitivities, predictions, and uncertainty: Wiley and Sons, New York, New York.

Hill, M.C. and Østerby, Ole, 2003, Determining extreme parameter correlation in ground-water models: Ground Water, v. 41, No. 4, p. 420-430.

Keidser, A. and Rosbjerg, D., 1991, A comparison of four inverse approaches to groundwater flow and transport parameter identification: Water Resources Research, v. 27, No. 9, p. 2219-2232.

Matott, L.S., 2005, OSTRICH, An optimization software tool, documentation and user's guide, Version 1.6: State University of New Yord at Buffolo, 114p. Accessed December 28, 2005 at http://www.groundwater.buffalo.edu/software/Ostrich/OstrichMain.html

Mehl, S.W. and Hill, M.C., 2002, Evaluation of a local grid refinement method for steady-state block-centered finite-difference groundwater models: p. 367-374 in S.M. Hassanizadeh, R.J. Schotting, W.G. Gray, and G.F. Pinder, eds., Proceedings of the XIVth International Conference on Computer Methods in Water Resources Conference, Developments in Water Science vol. 47, Elsevier, June, 2002, Delft, the Netherlands, ISBN: 0-444-50975-5, 1808 p.

Mehl, S.W. and Hill, M.C., 2003, Locally refined block-centered finite-difference groundwater models, Evaluation of parameter sensitivity and the consequences for inverse modelling and predictions: Karel Kovar, Hrkal Zbynek, eds., IAHS Publication 277, p. 227-232.

Poeter, E.P. and Anderson, D.R., 2005, Multi-model ranking and inference in ground-water modeling: Ground Water, vol. 43, no. 4, p. 597-605.

Poeter, E.P. and Hill, M.C., 1996, Unrealistic parameter estimates in inverse modeling: a problem or a benefit for model calibration?: Proceedings of the ModelCARE 96 Conference, Golden, CO, September 1996, International Association of Hydrological Sciences Publication no. 237, p. 277-285.

Poeter, E.P. and Hill, M.C., 1997, Inverse models: A necessary next step in groundwater modeling: Ground Water, v. 35, No. 2, p. 250-260.

Poeter, E.P. and Hill, M.C., 1998, Documentation of UCODE, a computer code for universal inverse modeling: U.S Geological Survey Water-Resources Investigations Report 98-4080, 122p. http://pubs.water.usgs.gov/wri984080/.

Saltelli, Andrea, Chan, Karen, and Scott, E. M., 2000, Sensitivity Analysis, John Wiley & Sons, NY, 475p.

Seber, G.A.F., and C.J. Wild, 1989, Nonlinear Regression, John Wiley & Sons, NY, 768p.

Tiedeman, C.R., Ely, D.M., Hill, M.C., and O'Brien, G.M., 2004, A method for evaluating the importance of system state observations to model predictions, with application to the Death Valley regional groundwater flow system: Water Resources. Research, v. 40, W12411, doi:10.1029/2004WR003313.

Tiedeman, C.R., Hill, M.C., D'Agnese, F.A., and Faunt, C.C., 2003, Methods for using groundwater model predictions to guide hydrogeologic data collection, with application to the Death Valley regional ground-water flow system: Water Resources Research, v. 39, No. 1, p. 5-1 to 5-17, 10.1029/2001WR001255.

Tonkin, M.J., Hill, M.C., and Doherty, John, 2003, MODFLOW-2000, the U.S. Geological Survey modular ground-water model -- Documentation of MOD-PREDICT for predictions, prediction sensitivity analysis, and enhanced analysis if model fit: U.S. Geological Survey Open-File Report 03-385, 69 p. http://water.usgs.gov/nrp/gwsoftware/modflow2000/modflow2000.html

Wagner B.J., and Gorelick, S.M., 1986. A statistical methodology for estimating transport parameters: theory and applications to one-dimensional advective-dispersive systems: Water Resources Research, v. 22, No. 8, p. 1303-1315.

Winston, R.B., 2000, Graphical User Interface for MODFLOW, Version 4: U.S. Geological Survey Open-File Report 00-315, 27 p. http://water.usgs.gov/nrp/gwsoftware/GW_Chart/GW_Chart.html

Yager, R.M., 1998, Detecting influential observations in nonlinear regression modeling of ground-water flow: Water Resources Research, v. 34, no. 7, p. 1623-1633.

Yager, R.M., 2004, Effects of model sensitivity and nonlinearity on nonlinear regression of ground water flow: Ground Water 42(3):390-400.

Appendix A. CONNECTION WITH THE JUPITER API

UCODE_2005, RESIDUAL_ANALYSIS, LINEAR_UNCERTAINTY, MODEL_LINEARITY, RESIDUAL_ANALYSIS_ADV, LINEAR_UNCERTAINTY_ADV, and CORFAC_PLUS were designed and constructed using conventions and tools from the Joint Universal Parameter IdenTification and Evaluation of Reliability (JUPITER) Application Programming Interface (API) (E.R. Banta, E.P. Poeter, M.C. Hill, and John Doherty, written commun., 2005).

The JUPITER API is a computer programming environment that includes conventions and software components designed to support the development of computer programs that perform model sensitivity analysis, data needs evaluation, calibration, uncertainty evaluation, and(or) optimization. The goal of the JUPITER API is to allow scientists to be able to express their ideas in programs that are sophisticated enough to be readily used in research and applications. For example, the JUPITER API provides modules that make it easy for such programs to use or expand existing input blocks, substitute parameter values into model input files, extract data from model output files, use full weight matrices, and produce data-exchange files.

It is hoped that facilitating the connection between ideas, research, and application in this way will accelerate technical and scientific advances related to analysis of natural systems, and, therefore, lead to more successful societal decisions about these systems.

This appendix describes the aspects of the API used for each program.

UCODE_2005

UCODE_2005 uses the JUPITER modules listed in Table A-1.

UCODE_2005 replaces UCODE (Poeter and Hill, 1998). To improve modularity, UCODE_2005 is now limited to performing sensitivity analysis and regression, while other stand-alone JUPITER applications are used to generate sets of normally distributed random numbers for residual analysis (RESIDUAL_ANALYSIS), calculate linearity (MODEL_LINEARITY) and evaluate predictive uncertainty (LINEAR_UNCERTAINTY). If desired, multiple programs can be run using a single file (such as a batch file on Windows operating systems), thereby achieving performance similar to that of UCODE.

UCODE was written using the computer languages Perl and Fortran77. UCODE_2005 uses only Fortran90/95. As part of this change, the substitution and reading processes used to interact with the process model(s), which had been written in Perl, were changed using Fortran code provided by John Doherty (written commun., 2003) as part of the JUPITER API Model Input-Output Modules. The result is that the methods used for substitution and reading are nearly identical to those used in PEST (Doherty, 2004), which makes it easier for users to compare and use the analyses provided by both of these programs. The only difference is that UCODE_2005 has a standard matrix option that is not available in the version of PEST documented by Doherty (2004).

The JUPITER API input and output conventions are also used by a version of PEST called J_PEST. This compatibility allows greater ease of code comparison and use of capabilities unique to UCODE_2005 and J_PEST.

Table A- 1. JUPITER modules, conventions, and mechanisms used in UCODE_2005. [The modules listed are described in Banta and others (written commun., 2005).]

Module	Authorship
Datatypes	Banta
Global Data	Banta and Doherty
Utilities	Banta, Doherty, and Poeter
Basic	Banta and Doherty
Model Input-Output	Doherty and Banta
Dependents	Banta and Poeter
Equation	Doherty
Sensitivity	Banta and Poeter
Statistics	Poeter, Hill, and Banta
Prior-Information	Banta and Poeter
Parallel Processing	Banta
Conventions and Mechanisms	**Authorship**
Parameter-Value Generation	Hill, Banta, Poeter, and Doherty
Input and Output Specifications (including matrices)	Banta, Doherty, Hill, and Poeter
Input Block Names and Keywords	Doherty, Poeter, and Banta
Data-Exchange Files	Poeter, Banta, and Hill

The conventions and some of the content of the input blocks and data-exchange files produced and used by the programs documented in this report are defined in the JUPITER API. For more information, see the JUPITER API documentation.

The Other Six Codes

The other six codes documented in this report are RESIDUAL_ANALYSIS, RESIDUAL_ANALYSIS_ADV, LINEAR_UNCERTAINTY, MODEL_LINEARITY, MODEL_LINEARITY_ADV, and CORFAC_PLUS. These programs use the input block and data exchange-file conventions established as part of the JUPITER API and JUPITER API modules as indicated in Table A-2.

Table A-2. JUPITER API modules, conventions, and mechanisms used in the other six codes documented in this report.
[The modules listed are described in Banta and others, written commun., 2005).

Module	Authorship
Datatypes	Banta
Global Data	Banta and Doherty
Utilities	Banta, Doherty, and Poeter
Basic	Banta and Doherty
Dependents	Banta and Poeter
Equation	Doherty
Sensitivity	Banta and Poeter
Statistics	Poeter, Hill, and Banta
Prior-Information	Banta and Poeter
Conventions and Mechanisms	**Authorship**
Input and Output Specifications (including matrices)	Banta, Doherty, Hill, and Poeter
Input Block Names and Keywords	Doherty, Poeter, and Banta
Data-Exchange Files	Poeter, Banta, and Hill

References

Doherty, John, 2004, PEST-2000: Corinda, Australia, Watermark Computing, http://www.sspa.com/PEST/index.html

Appendix A. Connection of UCODE_2005 with the JUPITER API

Appendix B: FILES PRODUCED BY USING THE FILENAME PREFIX SPECIFIED ON COMMAND LINES

The programs documented in this report produce a number of files using the fn filename prefix specified on the command line. File extensions for all of the files are listed in Table B- 1 in alphabetical order.

Table B- 1. Files produced by UCODE_2005, RESIDUAL_ANALYSIS, LINEAR_UNCERTAINTY, and MODEL_LINEARITY named using the fn prefix specified on the command line, in alphabetic order by letter in the file extension.

[File extensions that begin with an underscore identify data-exchange files and are listed first. File extensions that begin with # are main output files and are listed at the end of the table. Shading identifies files not produced by UCODE_2005. 'iteration' refers to parameter-estimation iteration. * is replaced by 'conf' for results related to confidence intervals or 'pred' for results related to prediction intervals.]

Extension	Brief description (see also tables 16-22, 24-28, 34, 35)	Content[1]	Use[1]
_b1	Parameter sets for _b2	19	--
_b1adv*	Parameter sets for _b2adv*	35	34
_b2	Values simulated using parameter sets from _b1	19	--
_b2adv*	Values simulated using parameter sets from _b1adv*	35	34
_b3*	Parameter sets for _b4*	35	34
_b4*	Values simulated using parameter sets from _b3*	35	34
_cf*	Correction factors	35	34
_cfsu	Sensitivities used by CORFAC_PLUS	35	34
_dm	Information about model structure, fit and parsimony	21	--
_dmp	Information about predictions	21	--
_gm	Observation groups	16	--
_gmp	Prediction groups	20	--
_init	As for _paopt evaluated at other parameter values	15	--
_init._mv	As for the _mv file evaluated at other parameter values	15	--
_init._su	As for the _su file evaluated at other parameter values	15	--
_init._supri	As for the _supri file evaluated at other parameter values	15	--
_int*	Nonlinear intervals	35	34
_int*par	Parameter values at nonlinear interval limits	35	34
_int*wr	Weighted residuals at nonlinear interval limits		
_linp	Predictions and their linear confidence intervals	26	--
#linunc	LINEAR_UNCERTAINTY main output file	26	--
_mc	Parameter correlation coefficient matrix	19	30

231

Appendix B: Files Produced by Using the Filename Prefix Specified on Command Lines

Extension	Brief description (see also tables 16-22, 24-28, 34, 35)	Content[1]	Use[1]
#modlin	MODEL_LINEARITY main output file	27	--
#modlinadv	MODEL_LINEARITY_ADV main output file		
_mv	Parameter variance-covariance matrix	19	30
_nm	Weighted residuals, probability plotting positions	16	31
_os	Unweighted simulated and observed or prior values	16	31
_p	Predictions	20	--
_pa	Parameter values for each iteration	19	--
_paopt [6]	Information for all defined parameters	19	--
_pasub	Parameter values for each iteration	19	--
_pc [6]	Information for estimated parameters	19	30
_pcc	Large parameter correlation coefficients (• 0.85)	18	30, 31
_pr	Prior information equations	18	--
_pv	Prediction variances	20	--
_r	Unweighted residuals (observations and prior)	16	31
_rb	DFBetas statistics for each observation and parameter	24	32
_rc	Cook's D statistic for each observation	24	32
_rd	Ordered uncorrelated random numbers	24	32
_rdadv	Results of RESIDUAL_ANALYSIS_ADV	--	--
_rg	Ordered correlated random numbers	24	32
_s1	One-percent scaled sensitivities	18	30, 31
_sc	Composite scaled sensitivity	18	30, 31
_sd	Dimensionless scaled sensitivities	18	30, 31
_so	Sensitivity summary by observation	18	30, 31
_sos	Parameter values and resulting value of the sum of squared weighted residuals objective function.	19	--
_sppp	Prediction sensitivity scaled by Param[5]/PredValue[4]	20	--
_sppr	Prediction sensitivity scaled by Param[5]/RefValue[3]	20	--
_spsp	Prediction sensitivity scaled by PSD[2]/PredValue[4]	20	--
_spsr	Prediction sensitivity scaled by PSD[2]/RefValue[3]	20	--
_spu	Unscaled sensitivities for predictions	20	--
_ss	Sum of squared weighted residuals by iteration	19	31
_su	Unscaled sensitivities for observations	18	--
_supri	Unscaled sensitivities for prior information	18	--
_w	Weighted residuals, observations and prior information	16	31
_ws	Simulated equivalents and weighted residuals for observations and prior information	16	31
_wt	Weights for observations	17	--
_wtpri	Weights for prior information	17	--

Appendix B: Files Produced by Using the Filename Prefix Specified on Command Lines

Extension	Brief description (see also tables 16-22, 24-28, 34, 35)	Content[1]	Use[1]
_ww	Weighted simulated equivalents in relation to weighted observations or prior information	16	31
_xyztwr	Merger of _r _w and the optional input file fn.xyzt	16	31
The following are main output files for the listed codes			
#corfac_*	CORFAC_PLUS	35	34
#linunc	LINEAR_UNCERTAINTY	26	--
#modlin	MODEL_LINEARITY	27	--
#modlinadv	MODEL_LINEARITY_ADV	--	--
#resan	RESIDUAL_ANALYSIS	24	--
#resanadv	RESIDUAL_ANALYSIS_ADV	--	--
The following are UCODE_2005 main output files for the listed modes			
#ucreateinitfiles	Sensitivity-analysis with CreateInitFiles=yes	--	--
#umodlin	Test-model-linearity	--	--
#unonlinint_*	Nonlinear-uncertainty	--	--
#uout	Forward, sensitivity-analysis, parameter-estimation	--	28-30
#upred	Prediction	20	33
#usos	Evaluate-objective-function	--	--

[1] Tables in this report that describe the file contents and suggest how to use it. --, the content or use of the file is not described in a table.

[2] Parameter Standard Deviation.

[3] Reference Value from Prediction_Data block. Scaled sensitivity is set to zero if this number equals zero.

[4] Predicted Value. Scaled sensitivity is set to zero if this number equals zero.

[5] Parameter Value.

[6] These data-exchange files are not generated from JUPITER API subroutines; they are part of UCODE_2005.

Appendix B: Files Produced by Using the Filename Prefix Specified on Command Lines

Appendix C: EXAMPLE SIMULATION

Examples are included in the distribution to demonstrate UCODE_2005 input and operation and to test the calculations. The directory structure and the batch files are described in appendix D. A complete set of input files is distributed with UCODE_2005. Output files can be produced by executing the batch files. The test case is referred to as test case 1 so that it can be differentiated easily from other test cases. This appendix describes the simulated system, discusses selected results, and presents selected input and output from the parameter-estimation mode run of UCODE_2005 (Table 3).

For the results presented here, test case 1 is simulated using the Ground-Water Flow Process of MODFLOW-2000 (Harbaugh and others, 2000; Hill and others, 2000; Anderman and Hill, 2001).

Simulations for calibration and predictions are presented.

Calibration Conditions

The physical system for test case 1 is shown in figure C-1. The system consists of two aquifers separated by a confining unit. Each aquifer is 50 m thick, and the confining unit is 10 m thick. The river is treated as a head-dependent boundary that is hydraulically connected to aquifer 1. Recharge from the hillside adjoining the system is treated as a head-dependent boundary that is hydraulically connected to aquifers 1 and 2 at the boundary farthest from the river.

Stresses on the system include (1) areal recharge to aquifer 1 in the area near the stream (zone 1) and in the area farther from the stream (zone 2), and (2) pumpage from wells completed in each of the two layers. Pumpage from aquifer 1 is assumed to equal pumpage from aquifer 2. The transient response to pumpage is simulated starting from a steady flow field with no pumpage.

Observations of head and river-flow gain are available for comparison with steady- and transient-state model results. The river is represented using MODFLOW-2000's River Package.

For the finite-difference method, the system is discretized into square 1,000-m by 1,000-m cells, so that the grid has 18 rows and 18 columns. Time discretization for the model run is specified to simulate a period of steady-state conditions with no pumpage followed by a transient-state period with a constant rate of pumpage. The steady-state period is simulated with one stress period having one time step. The transient period is simulated with four stress periods: the first three are 1, 3, and 6 days long, and each has one time step; the fourth is 272.8 days long and has 9 time steps, and each time-step length is 1.2 times the length of the previous time-step length.

The top model layer represents a system that is unconfined so that the saturated ground-water system is bounded on the top by a water table. However, this layer is simulated as

235

having constant thickness because the results are similar to the simulation with an unconfined layer and execution times are shorter. Indeed, this is an approximation that can be used to advantage in many ground-water systems, especially in the early stages of model development.

The parameters that define aquifer geometry are shown in figure C-1 and ground-water properties are listed in tables C-1 and C-2. The hydraulic conductivity of aquifer 2 increases with distance from the river. The variation is simulated using the multiplier-array capability of MODFLOW-2000. In this case, a multiplier array is defined to represent a step function and contains the value 1.0 in columns 1 and 2, 2.0 in columns 3 and 4, and so on to the value 9.0 in columns 17 and 18; this multiplier array is referenced in the definition of parameter HK_2 in the input file for the Layer Property Flow Package. For this test case, parameters SS_1 and SS_2 are defined such that their values are storage coefficients. SS_1 and SS_2 are divided by the aquifer thickness (using a multiplier array defined to be the inverse of the aquifer thickness) to produce the specific-storage values expected by the Layer Property Flow Package.

The river is simulated using the River Package to designate 18 river cells in column 1 of layer 1; the head in the river is 100 m. The conductance of the riverbed for each cell is calculated as $([L_{RB} \times W_{RB} / b_{RB}] \times K\_RB)$, where, for each cell, L_{RB} is the length of the river, W_{RB} is the width of the river, and b_{RB} is the thickness of the riverbed. K_RB is a parameter defined to be the hydraulic conductivity of the riverbed material, so that the quantity $[L_{RB} \times W_{RB} / b_{RB}]$ is listed as Condfact for each cell in the input file for the River Package (Harbaugh and others, 2000). For this system, $L_{RB} = 1,000$ m, $W_{RB} = 10$ m, and $b_{RB} = 10$ m at each river cell, so all Condfact values equal 1,000 m.

Ground-water flow into the system from the adjoining hillside is represented using the General-Head Boundary Package. Thirty-six general-head-boundary cells are specified in column 18 of layers 1 and 2, each having an external head of 350 m and a hydraulic conductance of $1x10^{-7}$ m^2/s.

Recharge in zone 1 (RCH_1) applies to cells in columns 1 through 9, recharge in zone 2 (RCH_2) applies to cells in columns 10 through 18. A multiplier array defined as a constant is referenced in the definitions of the recharge parameters to convert the recharge rates from units of cm/yr to m/s.

Pumpage is simulated using the Well Package. Wells are located at the center of the cells at row 9, column 10; there is one well in each layer and both wells have the same pumping rate. The parameter WELLS_TR specifies the pumping rate for each of the wells.

The parameter values estimated using observations with and without noise (errors) added to the observations are listed in tables C-1 and C-2, along with the starting and the true parameter values. Selected input and output files from the run with noisy observations are presented in the following sections.

Appendix C: Example Simulation

The results without noise (errors) added to the observations are presented in table C-2 to demonstrate that the regression estimated the true parameter values when unmodified simulated values from the synthetic system were used to generate the observations. This constitutes a test of the regression algorithm, and it can be seen that all parameter values were correctly estimated.

The example available in test subdirectory ex1 produces the results shown in table C-1 and additional results. It demonstrates a UCODE_2005 set up to execute a sequence of UCODE_2005 modes and associated runs of RESIDUAL_ANALSYSIS, LINEAR_UNCERTAINTY, and MODEL_LINEARITY. The predictions considered are the coordinates of a particle at 10, 50, and 200 years. Sensitivity-equation sensitivities produced by MODFLOW-2000 are used. The batch files distributed for the Windows operating system in the ex1 subdirectory are listed in Appendix D.

The example available in test subdirectory ex1-true produces the results shown in table C-2. It demonstrates UCODE_2005 set up to execute the parameter-estimation mode of UCODE_2005 using perturbation sensitivities.

Appendix C: Example Simulation

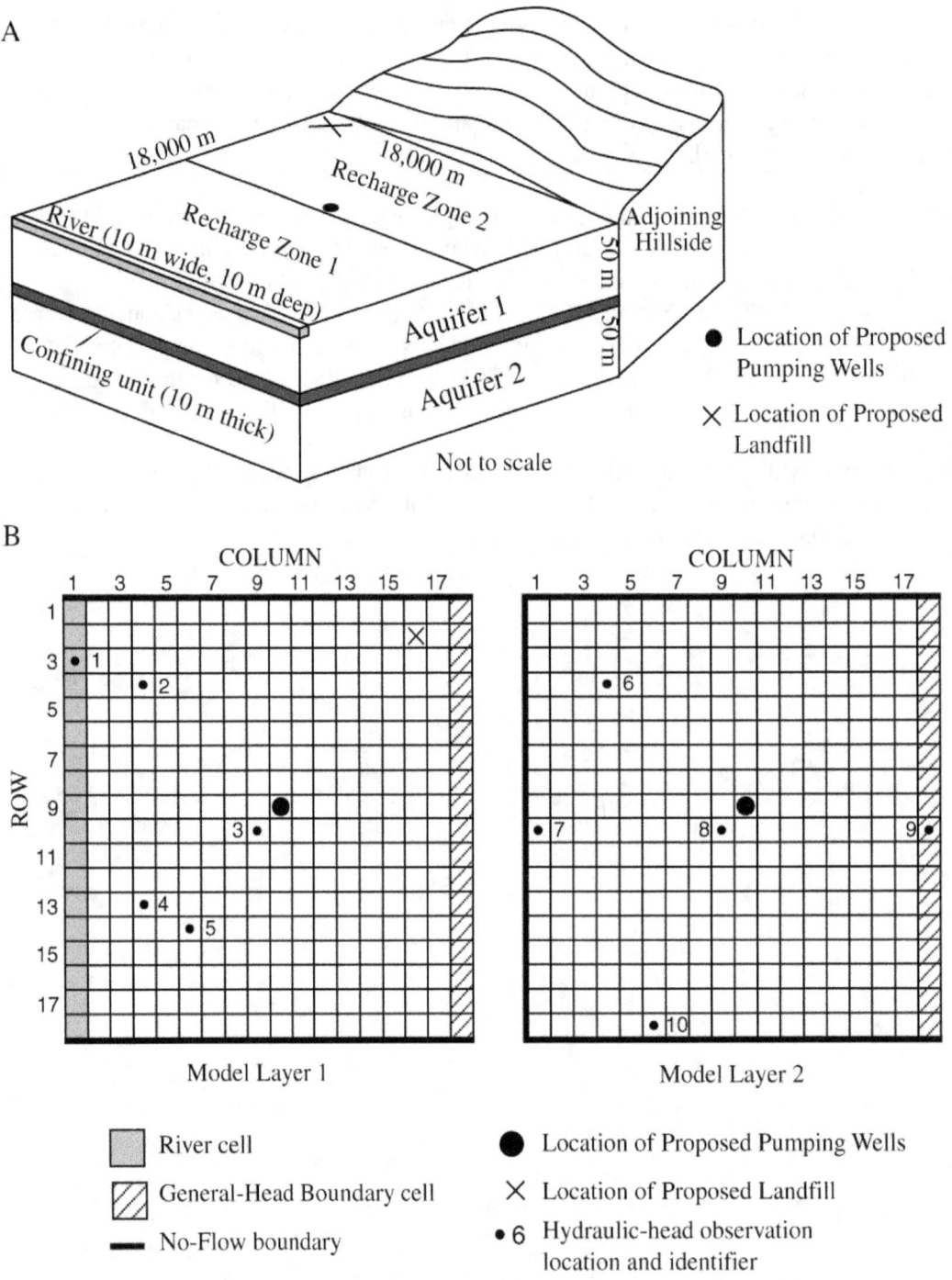

Figure C- 1: (A) Physical system and (B) model grid for test case 1. Pumpage is from two wells at the designated location. One pumps from aquifer 1, the other from aquifer 2.

Appendix C: Example Simulation

Table C- 1. Parameters defined for test case 1, starting and true parameter values, and the values estimated using the data with errors added.

[The associated output file main UCODE_2005 output file, ex1.#uout, is presented in this appendix. m, meter; s, second; cm, centimeter; yr, year. The estimated values and sum of squared, weighted residuals differ from those reported in Hill and others (2000) because variances on the drawdowns equal 0.0050 instead of 0.0025.]

Parameter label	Description	Starting Value	Estimated Value	True Value
Q_1&2	Pumping in each of two layers (m^3/s)	-1.10	-1.07	-1.00
RCH_1	Recharge rate in zone 1 (cm/yr)	60.0	34.1	31.6
RCH_2	Recharge rate in zone 2 (cm/yr)	30.0	50.5	47.3
K_RB	Hydraulic conductivity of the riverbed (m/s)	1.20×10^{-3}	1.38×10^{-3}	1.00×10^{-3}
SS_1	Storage coefficient of aquifer 1 (dimensionless)	1.30×10^{-3}	1.14×10^{-3}	1.00×10^{-3}
HK_1	Hydraulic conductivity of aquifer 1 (m/s)	3.00×10^{-4}	4.26×10^{-4}	4.00×10^{-4}
VK_CB	Vertical hydraulic conductivity of the confining layer (m/s)	1.00×10^{-7}	2.17×10^{-7}	2.00×10^{-7}
SS_2	Storage coefficient of aquifer 2 (dimensionless)	2.00×10^{-4}	6.09×10^{-5}	1.00×10^{-4}
HK_2	Hydraulic conductivity of aquifer 2 under the river (m/s)	4.00×10^{-5}	4.82×10^{-5}	4.40×10^{-5}
Sum of squared, weighted residuals (dimensionless)		269,000	23.9	171.7

Table C- 2: Parameters defined for test case 1, starting and true parameter values, and the values estimated using the data without errors added.
[This is from the run that can be executed from test directory "ex1-true" distributed with UCODE_2005. m, meter; s, second; cm, centimeter; yr, year.]

Parameter label	Description	Starting Values	Estimated Values	True Values
Q1&2	Pumping rate in each of layers 1 and 2 (m^3/s)	-1.10	-1.00	-1.00
RCH_1	Recharge rate in zone 1 (cm/yr)	60	31.6	31.6
RCH_2	Recharge rate in zone 2 (cm/yr)	30	47.3	47.3
K_RB	Hydraulic conductivity of the riverbed (m/s)	1.20×10^{-3}	1.00×10^{-3}	1.00×10^{-3}
SS_1	Storage coefficient of aquifer 1 (dimensionless)	1.30×10^{-3}	1.00×10^{-3}	1.00×10^{-3}
HK_1	Hydraulic conductivity of aquifer 1 (m/s)	3.00×10^{-4}	4.00×10^{-4}	4.00×10^{-4}
VK_CB	Vertical hydraulic conductivity of the confining layer (m/s)	1.00×10^{-7}	2.00×10^{-7}	2.00×10^{-7}
SS_2	Storage coefficient of aquifer 2 (dimensionless)	2.00×10^{-4}	1.00×10^{-4}	1.00×10^{-4}
HK_2	Hydraulic conductivity of aquifer 2 under the river (m/s)	4.00×10^{-5}	4.40×10^{-5}	4.40×10^{-5}
Sum of squared, weighted residuals (dimensionless)		268,770	1.33×10^{-6}	0.000000

Input Files

Selected input files are shown, including the UCODE_2005 main input file for estimating parameters using sensitivity-equation sensitivities, two files read by the Observation_Data input block, and two template files. The files are from the parameter-estimation mode run with noise in the observations.

UCODE_2005 Main Input File 03.in for Parameter-Estimation Mode

```
# -------------------------
# UCODE INPUT EXAMPLE 1
# -------------------------

BEGIN Options TABLE
NROW=1 NCOL=2 COLUMNLABELS
Verbose Derivatives_Interface
   0     ..\ex1a-files\transient.derint
END Options
```

Appendix C: Example Simulation
--UCODE_2005 Main Input File--

```
# -------------------------------
# UCODE-CONTROL INFORMATION
# ----------------------------

BEGIN UCODE_CONTROL_DATA KEYWORDS
ModelName=ex1
#Performance
  sensitivities=yes          # calculate sensitivities: yes/no
  optimize=yes               # estimate parameters: yes/no
#Printing and output files
  StartRes=no                # print residuals: yes/no
  IntermedRes=no             # # same
  FinalRes=no                # # same
  StartSens=css              # print sensitivities:
  IntermedSens=css           # #   css, dss, unscaled, onepercentss,
  FinalSens=css              # #   allss,all, or none
  DataExchange=yes           # create data-exchange files: yes/no
END UCODE_CONTROL_DATA

# ----------------------------
# REGRESSION-CONTROL INFORMATION
# ----------------------------

BEGIN REG_GN_CONTROLS KEYWORDS
tolpar=0.01                # GN param conv crit. Also see parameter blocks
tolsosc=0.0                # GN fit-change conv criteria.
MrqtDirection=85.41        # angle (degrees) for Mrqt parameter
MrqtFactor=1.5             # #
MrqtIncrement=0.001        # #
quasinewton=no             # option to use QN updating: yes, no
FletcherReeves=0           # # FR iterations for FR, QN combined
maxiter=10                 # maximum # of GaussNewton updates
maxchange=2.0              # max frac param change for GN updates
maxchangerealm=regression  # how changes apply, log-trans params
END REG_GN_CONTROLS

# --------------------------------
# COMMAND FOR APPLICATION MODEL(S)
# --------------------------------

BEGIN MODEL_COMMAND_LINES FILES
..\ex1a-files\obs-fwd.command
..\ex1a-files\obs-fwd-der.command
END MODEL_COMMAND_LINES
```

Appendix C: Example Simulation
--UCODE_2005 Main Input File--

```
# --------------------
# PARAMETER INFORMATION
# --------------------

BEGIN PARAMETER_GROUPS KEYWORDS
    GroupName = MyPars  adjustable=yes TOLPAR=.01  maxchange=2.0
    SENMETHOD=-1
END PARAMETER_GROUPS

BEGIN PARAMETER_DATA FILES
..\ex1a-files\tr.params
END PARAMETER_DATA

# ----------------------
# OBSERVATION INFORMATION
# ----------------------

BEGIN OBSERVATION_GROUPS FILES
..\ex1a-files\groups.obs
END OBSERVATION_GROUPS

BEGIN OBSERVATION_DATA FILES
..\ex1a-files\hed.obs
..\ex1a-files\flo.obs
END OBSERVATION_DATA

# ----------------------------
# APPLICATION MODEL INFORMATION
# ----------------------------

BEGIN MODEL_INPUT_FILES KEYWORDS
  modinfile=..\..\test-data-win\data-transient\tc1-fwd-der.sen
  templatefile=..\ex1a-files\tc1sen-eq.tpl
END MODEL_INPUT_FILES

BEGIN MODEL_OUTPUT_FILES  KEYWORDS
  modoutfile=..\..\test-data-win\data-transient\tc1._os
  instructionfile=..\ex1a-files\obs.instructions
  category=obs
END MODEL_OUTPUT_FILES
```

Other Selected UCODE_2005 Input Files

File Listed in the Observation_Data Input Block: flow.obs

```
BEGIN OBSERVATION_DATA TABLE
NROW=3  NCOL=4  COLUMNLABELS  GROUPNAME=FLOWS
obsname          obsvalue  statistic  statflag
flow01.ss          -4.4     .160000    var
flow01.10          -4.1     .144400    var
flow01.283         -2.2     .044100    var
END OBSERVATION_DATA
```

Template File Listed in the Model_Input_Files input block: tc1sen-eq.tpl

```
jtf @
    9    0   -40    9        ITEM 1: NPLIST ISENALL IUHEAD MXSEN
    0    0    0     0        ITEM 2: IPRINTS ISENSU ISENPU ISENFM
Q1&2          1  0 @Q1&2     @    -1.4    -0.8   1.0E-3
RCH_1         1  0 @RCH_1    @    30.0    80.0   1.0E-2
RCH_2         1  0 @RCH_2    @    20.0    60.0   1.0E-2
K_RB          1  1 @K_RB     @   1.2E-4  1.2E-2  1.0E-6
SS_1          1  1 @SS_1     @   1.3E-4  1.3E-2  1.0E-6
HK_1          1  1 @HK_1     @   3.0E-5  3.0E-3  1.0E-7
VK_CB         1  1 @VK_CB    @   1.0E-8  1.0E-6  1.0E-10
SS_2          1  1 @SS_2     @   2.0E-5  2.0E-3  1.0E-7
HK_2          1  1 @HK_2     @   4.0E-6  4.0E-4  1.0E-8
```

Output Files

Selections from the main output file and selected data-exchange files are shown from the UCODE_2005 parameter-estimation mode using sensitivity-equation sensitivities. Some lines have been editing so they do not wrap or omitted to improve presentation.

UCODE_2005 Main output file ex1.#03uout-parest

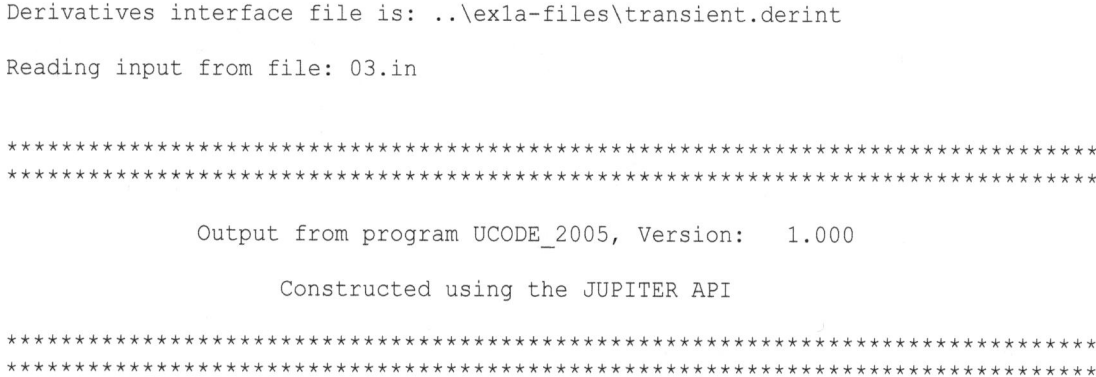

```
    Derivatives interface file is: ..\ex1a-files\transient.derint

    Reading input from file: 03.in


    *********************************************************************
    *********************************************************************


            Output from program UCODE_2005, Version:   1.000

            Constructed using the JUPITER API


    *********************************************************************
    *********************************************************************
```

Appendix C: Example Simulation
--UCODE_2005 Main Output File--

```
ECHO UCODE CONTROLS INPUT:

MODEL NAME =   ex1
MODEL LENGTH UNITS =   NA
MODEL TIME UNITS =   NA
MODEL MASS UNITS =   NA


OPTIMIZATION WILL BE PERFORMED
NONLINEAR INTERVALS WILL NOT BE PERFORMED (BY DEFAULT)
SENSITIVITIES WILL BE CALCULATED
STARTRES - Residuals will not be written at initial parameter values
INTERMEDRES - Residuals will not be written between iterations
FINALRES - Residuals will not be written at final parameter values
EIGENVECTORS/EIGENVALUES WILL NOT BE PRINTED
STARTSENS Print composite scaled sensitivities at starting values
INTERMEDSENS Print composite scaled sensitivities at intermed values
FINALSENS Print composite scaled sensitivities at final values
DATA EXCHANGE FILES WILL BE PRODUCED


ECHO MODIFIED GAUSS-NEWTON UCODE REGRESSION CONTROL INPUT:

MAXIMUM NUMBER OF PARAMETER ITERATIONS            =           10
DEFAULT TOLERANCE ON % PARAMETER CHANGE FOR CLOSURE  =    1.00E-02
TOLERANCE ON % CHANGE OF SOSC OVER 3 ITERATIONS   =    0.00E+00
MAXIMUM FRACTIONAL PARAMETER CHANGE               =    2.00E+00
MAXIMUM CHANGE APPLIES TO INDICATED PARAMETER SPACE    REGRESSION
MARQUARDT DIRECTION (IN DEGREES)                  =    85.41
MARQUARDT FACTOR                                  =    1.50E+00
MARQUARDT INCREMENT                               =    1.00E-03


ECHO MODEL COMMAND LINES:

Command ID    Purpose       Command Line
------------  ------------  --------------------------------------------------
modflow       Forward       ..\..\test-data-win\data-transient\tc1-fwd.bat
modflow_sen   Forward&Der   ..\..\test-data-win\data-transient\tc1-fwd-der.bat

PARAMETER INFORMATION:

No.  Param. name  Group       Value         Lower value    Upper value    Adj?
---- ------------ ----------- ------------  ------------   ------------   -----
   1 Q1&2         MyPars      -1.10000      -1.40000       -0.800000       Y
   2 RCH_1        MyPars      60.0000       30.0000        80.0000         Y
   3 RCH_2        MyPars      30.0000       20.0000        60.0000         Y
   4 K_RB         MyPars      1.200000E-03  1.200000E-04   1.200000E-02    Y
   5 SS_1         MyPars      1.300000E-03  1.300000E-04   1.300000E-02    Y
   6 HK_1         MyPars      3.000000E-04  3.000000E-05   3.000000E-03    Y
   7 VK_CB        MyPars      1.000000E-07  1.000000E-08   1.000000E-06    Y
   8 SS_2         MyPars      2.000000E-04  2.000000E-05   2.000000E-03    Y
   9 HK_2         MyPars      4.000000E-05  4.000000E-06   4.000000E-04    Y

No.  Param. name  LN  BSCAL      PERTURB     MAXCHANGE   TOLPAR
---- ------------ --  ---------- ----------  ---------   ---------
   1 Q1&2         0   1.000E-03  1.000E-02   2.00        1.000E-02
   2 RCH_1        0   1.000E-02  1.000E-02   2.00        1.000E-02
   3 RCH_2        0   1.000E-02  1.000E-02   2.00        1.000E-02
   4 K_RB         1   1.000E-06  1.000E-02   2.00        1.000E-02
```

```
        5 SS_1            1    1.000E-06   1.000E-02   2.00      1.000E-02
        6 HK_1            1    1.000E-07   1.000E-02   2.00      1.000E-02
        7 VK_CB           1    1.000E-10   1.000E-02   2.00      1.000E-02
        8 SS_2            1    1.000E-07   1.000E-02   2.00      1.000E-02
        9 HK_2            1    1.000E-08   1.000E-02   2.00      1.000E-02
```

INFORMATION FOR ADJUSTABLE PARAMETERS:

Param. name	Group	Value	Lower value	Upper value	Par. no.
Q1&2	MyPars	-1.10000	-1.40000	-0.800000	1
RCH_1	MyPars	60.0000	30.0000	80.0000	2
RCH_2	MyPars	30.0000	20.0000	60.0000	3
K_RB	MyPars	1.200000E-03	1.200000E-04	1.200000E-02	4
SS_1	MyPars	1.300000E-03	1.300000E-04	1.300000E-02	5
HK_1	MyPars	3.000000E-04	3.000000E-05	3.000000E-03	6
VK_CB	MyPars	1.000000E-07	1.000000E-08	1.000000E-06	7
SS_2	MyPars	2.000000E-04	2.000000E-05	2.000000E-03	8
HK_2	MyPars	4.000000E-05	4.000000E-06	4.000000E-04	9

Param. name	Sens Method	LN	BSCAL	PERTURB	MAXCHANGE	TOLPAR
Q1&2	-1	0	1.000E-03	1.000E-02	2.00	1.000E-02
RCH_1	-1	0	1.000E-02	1.000E-02	2.00	1.000E-02
RCH_2	-1	0	1.000E-02	1.000E-02	2.00	1.000E-02
K_RB	-1	1	1.000E-06	1.000E-02	2.00	1.000E-02
SS_1	-1	1	1.000E-06	1.000E-02	2.00	1.000E-02
HK_1	-1	1	1.000E-07	1.000E-02	2.00	1.000E-02
VK_CB	-1	1	1.000E-10	1.000E-02	2.00	1.000E-02
SS_2	-1	1	1.000E-07	1.000E-02	2.00	1.000E-02
HK_2	-1	1	1.000E-08	1.000E-02	2.00	1.000E-02

```
OBSERVATIONS
  Total number of observations read---------    35
  Number of directly extracted observations-    35
  Number of observations to be derived------     0
  Number of observations to be used---------    35


Number of linear prior-information equations =     0

*************************************************************************
END ECHO OF INPUT - REPORT RESULTS OF SIMULATION
*************************************************************************


Method of obtaining sensitivities is: MODEL-CALCULATED SENSITIVITY
```

```
------------------------------------------------------------------
FIT OF SIMULATED EQUIVALENTS TO OBSERVATIONS
```

OBSERVATION NAME	MEASURED VALUE	SIMULATED VALUE	RESIDUAL	WEIGHT**.5	WEIGHTED RESIDUAL
hd01.ss	101.804	100.214	1.590	0.999	1.588
hd01.1	-2.900000E-02	-1.525879E-05	-2.8985E-02	14.1	-0.4099
hd01.283	-0.129000	-9.061432E-02	-3.8386E-02	14.1	-0.5429
hd02.ss	128.117	137.416	-9.299	0.999	-9.287
hd02.1	-4.100000E-02	-9.475708E-03	-3.1524E-02	14.1	-0.4458
hd02.4	-0.557000	-0.276215	-0.2808	14.1	-3.971
hd02.10	-11.5310	-12.9634	1.432	14.1	20.26
hd02.283	-14.1840	-18.7714	4.587	14.1	64.88
hd03.ss	156.678	170.742	-14.06	0.999	-14.05
hd03.1	-4.38100	-3.66571	-0.7153	14.1	-10.12
hd03.283	-42.5400	-56.2375	13.70	14.1	193.7
hd04.ss	124.893	137.416	-12.52	0.999	-12.51
hd04.1	-6.700000E-02	-1.625061E-02	-5.0749E-02	14.1	-0.7177
hd04.283	-14.3040	-18.8487	4.545	14.1	64.27
hd05.ss	140.961	154.350	-13.39	0.999	-13.37
hd05.1	-6.000000E-02	-3.678894E-02	-2.3211E-02	14.1	-0.3283
hd05.283	-21.6760	-28.4615	6.785	14.1	95.96
hd06.ss	126.537	137.702	-11.17	0.999	-11.15
hd06.1	5.000000E-03	-1.245117E-02	1.7451E-02	14.1	0.2468
hd06.283	-14.3650	-19.1850	4.820	14.1	68.17
hd07.ss	101.112	102.728	-1.616	0.999	-1.614
hd07.1	4.800000E-02	-1.144409E-03	4.9144E-02	14.1	0.6950
hd07.283	-0.568000	-1.38068	0.8127	14.1	11.49
hd08.ss	158.135	170.355	-12.22	0.999	-12.21
hd08.1	-5.53300	-5.80962	0.2766	14.1	3.912
hd08.283	-43.2170	-57.2549	14.04	14.1	198.5
hd09.ss	176.374	185.904	-9.530	0.999	-9.518
hd09.1	-1.000000E-03	-5.050659E-02	4.9507E-02	14.1	0.7001
hd09.283	-38.2420	-49.5115	11.27	14.1	159.4
hd10.ss	142.020	154.263	-12.24	0.999	-12.23
hd10.1	-1.300000E-02	-4.287720E-03	-8.7123E-03	14.1	-0.1232
hd10.283	-19.9210	-26.1288	6.208	14.1	87.79
flow01.ss	-4.40000	-4.62400	0.2240	2.50	0.5600
flow01.10	-4.10000	-4.48159	0.3816	2.63	1.004
flow01.283	-2.20000	-2.62660	0.4266	4.76	2.031

```
STATISTICS FOR ALL RESIDUALS :
AVERAGE WEIGHTED RESIDUAL  : 0.246E+02
# RESIDUALS >= 0. :      18
# RESIDUALS < 0.  :      17
NUMBER OF RUNS  :   19  IN   35 OBSERVATIONS


THE NUMBER OF RUNS EQUALS THE EXPECTED NUMBER OF RUNS


STATISTICS FOR THESE RESIDUALS:
MAXIMUM WEIGHTED RESIDUAL: 0.199E+03  Observation: hd08.283
MINIMUM WEIGHTED RESIDUAL: -0.140E+02  Observation: hd03.ss
AVERAGE WEIGHTED RESIDUAL: 0.246E+02
# RESIDUALS >= 0. :      18
# RESIDUALS < 0.  :      17
NUMBER OF RUNS:      19  IN        35 OBSERVATIONS
```

246

```
SUM OF SQUARED WEIGHTED RESIDUALS            :    0.13406E+06

SUM OF SQUARED WEIGHTED RESIDUALS WITH PRIOR:    0.13406E+06

SUM OF SQUARED, WEIGHTED RESIDUALS:
  DEPENDENT VARIABLES:  0.13406E+06

NUMBER OF INCLUDED OBSERVATIONS =        35 OF        35

*****************************************************************************
*****************************************************************************

CALCULATING SENSITIVITIES FOR THE INITIAL PARAMETERS

Method of obtaining sensitivities is: MODEL-CALCULATED SENSITIVITY

-----------------------------------------------------------------------------
-----------------------------------------------------------------------------

COMPOSITE SCALED SENSITIVITIES ((SUM OF THE SQUARED DSS)/ND)**.5
  DSS = DIMENSIONLESS SCALED SENSITIVITIES (SCALED BY (PARAMETER_VALUE*(wt**.5))

   PARAMETER      COMPOSITE SCALED SENSITIVITY    RATIO TO MAXIMUM
   ---------      ----------------------------    ----------------
   Q1&2                    2.58328E+02                1.00000E+00
   RCH_1                   1.49116E+01                5.77235E-02
   RCH_2                   1.37221E+01                5.31190E-02
   K_RB                    6.39961E-01                2.47732E-03
   SS_1                    4.42764E+01                1.71396E-01
   HK_1                    1.58725E+02                6.14433E-01
   VK_CB                   5.25642E+00                2.03479E-02
   SS_2                    7.58973E+00                2.93802E-02
   HK_2                    5.33679E+01                2.06590E-01

STARTING VALUES OF REGRESSION PARAMETERS :

Q1&2         RCH_1         RCH_2        K_RB         SS_1         HK_1
VK_CB        SS_2          HK_2


  -1.100        60.00         30.00      1.2000E-03   1.3000E-03   3.0000E-04
   1.0000E-07   2.0000E-04    4.0000E-05

=============================================================================
=============================================================================

UCODE Modified Gauss-Newton: Parameter-Estimation Iteration#:     1

VALUES FROM SOLVING THE NORMAL EQUATION :
  MRQT PARAMETER ------------------ = 0.0000
  FRACTIONAL PARAMETER CHANGE IS EVALUATED IN              REGRESSION SPACE
      MAXIMUM FRACTIONAL CHANGE OCCURRED FOR PARAMETER: "RCH_2         "
                  MAXIMUM FRACTIONAL PARAMETER CHANGE    = 0.465
          CONVERGENCE TOLERANCE FOR THIS PARAMETER (TolPar)    = 1.000E-02
          MAXIMUM CHANGE ALLOWED FOR THIS PARAMETER (MaxChange) =  2.00

ADJUSTMENTS TO PARAMETER CHANGE VECTOR WERE NOT REQUIRED
```

```
UPDATED ESTIMATES OF REGRESSION PARAMETERS :

Q1&2          RCH_1         RCH_2          K_RB          SS_1          HK_1
VK_CB         SS_2          HK_2

  -1.019         39.39         43.95       2.0469E-04    1.2419E-03    4.0013E-04
  1.9020E-07    8.8500E-05    4.3463E-05

Method of obtaining sensitivities is: MODEL-CALCULATED SENSITIVITY

SUM OF SQUARED, WEIGHTED RESIDUALS:
  DEPENDENT VARIABLES:    672.17

NUMBER OF INCLUDED OBSERVATIONS =       35 OF       35

*************************************************************************************
*************************************************************************************

CALCULATING SENSITIVITIES FOR PARAMETERS ESTIMATED IN ITERATION:       1

Method of obtaining sensitivities is: MODEL-CALCULATED SENSITIVITY

-------------------------------------------------------------------------------------
-------------------------------------------------------------------------------------

COMPOSITE SCALED SENSITIVITIES ((SUM OF THE SQUARED DSS)/ND)**.5
  DSS = DIMENSIONLESS SCALED SENSITIVITIES (SCALED BY (PARAMETER_VALUE*(wt**.5))

   PARAMETER    COMPOSITE SCALED SENSITIVITY    RATIO TO MAXIMUM
   ---------    ----------------------------    ----------------
   Q1&2                 2.00343E+02                1.00000E+00
   RCH_1                7.91070E+00                3.94858E-02
   RCH_2                1.60788E+01                8.02566E-02
   K_RB                 3.90115E+00                1.94724E-02
   SS_1                 2.31102E+01                1.15353E-01
   HK_1                 1.39173E+02                6.94673E-01
   VK_CB                3.43326E+00                1.71369E-02
   SS_2                 1.96707E+00                9.81849E-03
   HK_2                 3.88127E+01                1.93731E-01


=====================================================================================
=====================================================================================

UCODE Modified Gauss-Newton: Parameter-Estimation Iteration#:      2

VALUES FROM SOLVING THE NORMAL EQUATION :
  MRQT PARAMETER ------------------ =  0.0000
  FRACTIONAL PARAMETER CHANGE IS EVALUATED IN                    REGRESSION SPACE
      MAXIMUM FRACTIONAL CHANGE OCCURRED FOR PARAMETER:  "RCH_2          "
               MAXIMUM FRACTIONAL PARAMETER CHANGE       =  0.129
        CONVERGENCE TOLERANCE FOR THIS PARAMETER (TolPar) =  1.000E-02
        MAXIMUM CHANGE ALLOWED FOR THIS PARAMETER (MaxChange) =  2.00

ADJUSTMENTS TO PARAMETER CHANGE VECTOR WERE NOT REQUIRED
```

.

Lines omitted

```
================================================================================
================================================================================


   UCODE Modified Gauss-Newton: Parameter-Estimation Iteration#:    5

   VALUES FROM SOLVING THE NORMAL EQUATION :
     MRQT PARAMETER ------------------- =  0.0000
     FRACTIONAL PARAMETER CHANGE IS EVALUATED IN              REGRESSION SPACE
         MAXIMUM FRACTIONAL CHANGE OCCURRED FOR PARAMETER:  "K_RB        "
                       MAXIMUM FRACTIONAL PARAMETER CHANGE    =  8.148E-03
         CONVERGENCE TOLERANCE FOR THIS PARAMETER (TolPar)    =  1.000E-02
         MAXIMUM CHANGE ALLOWED FOR THIS PARAMETER (MaxChange) =  2.00

   ADJUSTMENTS TO PARAMETER CHANGE VECTOR WERE NOT REQUIRED

   UPDATED ESTIMATES OF REGRESSION PARAMETERS :

   Q1&2          RCH_1         RCH_2         K_RB          SS_1          HK_1
   VK_CB         SS_2          HK_2


    -1.074         34.13         50.49      1.3168E-03    1.1405E-03    4.2579E-04
     2.1745E-07   6.0338E-05    4.8202E-05

 ******************************************************************************
   Parameter Estimation CONVERGED: % change of PARAMETER VALUES less than TolPar
 ******************************************************************************

 Method of obtaining sensitivities is: MODEL-CALCULATED SENSITIVITY

 SEARCHING FOR ITERATION WITH LOWEST SUM OF SQUARED RESIDUALS


 PARAMETER VALUES FROM ITERATION        4
 YIELDED THE LOWEST SUM OF SQUARED RESIDUALS

 THEIR SUM OF SQUARED RESIDUALS IS NOTED BELOW
   Q1&2          RCH_1         RCH_2         K_RB          SS_1          HK_1
   VK_CB         SS_2          HK_2


    -1.074         34.13         50.48      1.2469E-03    1.1405E-03    4.2577E-04
     2.1746E-07   6.0262E-05    4.8198E-05

 ******************************************************************************
 ******************************************************************************

   SUM OF SQUARED, WEIGHTED RESIDUALS:
     DEPENDENT VARIABLES:   23.908

   NUMBER OF INCLUDED OBSERVATIONS =        35 OF        35

 ******************************************************************************
 ******************************************************************************

 CALCULATING SENSITIVITIES FOR THE FINAL PARAMETERS

 Method of obtaining sensitivities is: MODEL-CALCULATED SENSITIVITY
```

```
--------------------------------------------------------------------------------

   COMPOSITE SCALED SENSITIVITIES ((SUM OF THE SQUARED DSS)/ND)**.5
    DSS = DIMENSIONLESS SCALED SENSITIVITIES (SCALED BY (PARAMETER_VALUE*(wt**.5))

     PARAMETER    COMPOSITE SCALED SENSITIVITY    RATIO TO MAXIMUM
     ----------   ----------------------------    ----------------
     Q1&2                 1.97188E+02              1.00000E+00
     RCH_1                6.26392E+00              3.17662E-02
     RCH_2                1.69904E+01              8.61635E-02
     K_RB                 7.04247E-01              3.57144E-03
     SS_1                 1.81815E+01              9.22039E-02
     HK_1                 1.42921E+02              7.24793E-01
     VK_CB                3.23459E+00              1.64035E-02
     SS_2                 1.21711E+00              6.17232E-03
     HK_2                 4.11078E+01              2.08470E-01

  SMALLEST AND LARGEST WEIGHTED RESIDUALS

                 SMALLEST WEIGHTED RESIDUALS
                      WEIGHTED      PERCENT OF
     NAME             RESIDUAL      OBJ FUNC
     hd04.ss            -2.11         18.61
     hd08.283           -1.51          9.48
     hd03.1             -0.860         3.09
     hd06.ss            -0.677         1.92
     hd04.283           -0.676         1.91

                 LARGEST   WEIGHTED RESIDUALS
                      WEIGHTED      PERCENT OF
     NAME             RESIDUAL      OBJ FUNC
     hd09.283            1.65         11.39
     hd01.ss             1.61         10.82
     hd09.1              1.40          8.23
     hd07.283            1.25          6.55
     hd02.ss             1.11          5.16

  CORRELATION BETWEEN ORDERED WEIGHTED RESIDUALS AND NORMAL ORDER STATISTICS
  FOR OBSERVATIONS =      0.969


--------------------------------------------------------------------------------
  COMMENTS ON THE INTERPRETATION OF THE CORRELATION BETWEEN
  WEIGHTED RESIDUALS AND NORMAL ORDER STATISTICS:

  The critical value for correlation at the 5% significance level is 0.943

  IF the reported CORRELATION is GREATER than the 5% critical value, ACCEPT
  the hypothesis that the weighted residuals are INDEPENDENT AND NORMALLY
  DISTRIBUTED at the 5% significance level.  The probability that this
  conclusion is wrong is less than 5%.

  IF the reported correlation IS LESS THAN the 5% critical value REJECT the
  hypothesis that the weighted residuals are INDEPENDENT AND NORMALLY
  DISTRIBUTED at the 5% significance level.

  The analysis can also be done using the 10% significance level.
  The associated critical value is 0.952
```

Appendix C: Example Simulation
--UCODE_2005 Main Output File--

```
    ----------------------------------------------------------------------
            VARIANCE-COVARIANCE MATRIX FOR THE PARAMETERS
            -------------------------------------------


            Q1&2         RCH_1        RCH_2        K_RB         SS_1 ...
    ......................................................................
    Q1&2    3.92290E-03 -2.85537E-02 -0.22920     -3.45211E-03 -3.73680E-03
    RCH_1  -2.85537E-02  9.4525      -3.1418        6.52706E-03  2.42110E-02
    RCH_2  -0.22920     -3.1418      16.140         0.23733      0.21512
    K_RB   -3.45211E-03  6.52706E-03  0.23733       0.36739     -9.55763E-03
    SS_1   -3.73680E-03  2.42110E-02  0.21512      -9.55763E-03  6.47401E-03
    HK_1   -3.63304E-03  2.81370E-02  0.21154      -2.72208E-06  3.43245E-03
    VK_CB  -3.72298E-03  2.87034E-02  0.21051      -2.82153E-02  7.90985E-03
    SS_2   -2.66023E-03  4.51227E-02  0.20071       0.20043     -3.88368E-02
    HK_2   -3.70521E-03  2.16934E-02  0.21924       1.10397E-02  3.48186E-03


            ----------------------------------------
            CORRELATION MATRIX FOR THE PARAMETERS
            ----------------------------------------


            Q1&2         RCH_1        RCH_2        K_RB         SS_1 ...
    ......................................................................
    Q1&2    1.0000      -0.14828     -0.91088      -9.09325E-02 -0.74150
    RCH_1  -0.14828      1.0000      -0.25437       3.50253E-03  9.78707E-02
    RCH_2  -0.91088     -0.25437      1.0000        9.74646E-02  0.66550
    K_RB   -9.09325E-02  3.50253E-03  9.74646E-02   1.0000      -0.19598
    SS_1   -0.74150      9.78707E-02  0.66550      -0.19598      1.0000
    HK_1   -0.99192      0.15650      0.90041      -7.67976E-05  0.72950
    VK_CB  -0.49205      7.72828E-02  0.43375      -0.38534      0.81377
    SS_2   -5.35295E-02  1.84969E-02  6.29653E-02   0.41676     -0.60832
    HK_2   -0.94709      0.11296      0.87366       0.29159      0.69279

    THE CORRELATION OF THE FOLLOWING PARAMETER PAIRS >= .95
        PARAMETER    PARAMETER    CORRELATION
        Q1&2         HK_1             -0.99

    THE CORRELATION OF THE FOLLOWING PARAMETER PAIRS IS BETWEEN .90 AND .95
        PARAMETER    PARAMETER    CORRELATION
        Q1&2         RCH_2            -0.91
        Q1&2         HK_2             -0.95
        RCH_2        HK_1              0.90
        HK_1         HK_2              0.90

    THE CORRELATION OF THE FOLLOWING PARAMETER PAIRS IS BETWEEN .85 AND .90
        PARAMETER    PARAMETER    CORRELATION
        RCH_2        HK_2              0.87


CORRELATIONS GREATER THAN 0.95 COULD INDICATE THAT THERE MAY NOT BE ENOUGH
INFORMATION IN THE OBSERVATIONS AND PRIOR USED IN THE REGRESSION TO ESTIMATE
PARAMETER VALUES INDIVIDUALLY.
TO CHECK THIS, START THE REGRESSION FROM SETS OF INITIAL PARAMETER VALUES
THAT DIFFER BY MORE THAT TWO STANDARD DEVIATIONS FROM THE ESTIMATED
VALUES.  IF THE RESULTING ESTIMATES ARE WELL WITHIN ONE STANDARD DEVIATION
OF THE PREVIOUSLY ESTIMATED VALUE, THE ESTIMATES ARE PROBABLY
DETERMINED INDEPENDENTLY WITH THE OBSERVATIONS AND PRIOR USED IN
THE REGRESSION.  OTHERWISE, YOU MAY ONLY BE ESTIMATING THE RATIO
OR SUM OF THE HIGHLY CORRELATED PARAMETERS.
```

```
_____

PARAMETER SUMMARY
_____


_____

PARAMETER VALUES IN "REGRESSION" SPACE --- LOG TRANSFORMED AS APPLICABLE
_____

PARAMETER:        Q1&2       RCH_1      RCH_2      K_RB       SS_1
* = LOG TRNS:                                       *          *

UPPER 95% C.I.   -9.45E-01   4.04E+01   5.87E+01  -2.36E+00  -2.87E+00
FINAL VALUES     -1.07E+00   3.41E+01   5.05E+01  -2.90E+00  -2.94E+00
LOWER 95% C.I.   -1.20E+00   2.78E+01   4.22E+01  -3.45E+00  -3.01E+00

STD. DEV.         6.26E-02   3.07E+00   4.02E+00   2.63E-01   3.49E-02

COEF. OF VAR. (STD. DEV. / FINAL VALUE); "--" IF FINAL VALUE = 0.0
                  5.83E-02   9.01E-02   7.96E-02   9.06E-02   1.19E-02

_____

PARAMETER VALUES IN "REGRESSION" SPACE --- LOG TRANSFORMED AS APPLICABLE
_____

PARAMETER:        HK_1       VK_CB      SS_2       HK_2
* = LOG TRNS:      *           *          *          *

UPPER 95% C.I.   -3.32E+00  -6.55E+00  -3.51E+00  -4.26E+00
FINAL VALUES     -3.37E+00  -6.66E+00  -4.22E+00  -4.32E+00
LOWER 95% C.I.   -3.42E+00  -6.77E+00  -4.93E+00  -4.37E+00

STD. DEV.         2.54E-02   5.25E-02   3.45E-01   2.71E-02

COEF. OF VAR. (STD. DEV. / FINAL VALUE); "--" IF FINAL VALUE = 0.0
                  7.53E-03   7.87E-03   8.17E-02   6.28E-03

_____

PHYSICAL PARAMETER VALUES --- EXP10 OF LOG TRANSFORMED PARAMETERS
_____

PARAMETER:        Q1&2       RCH_1      RCH_2      K_RB       SS_1
* = LOG TRNS:                                       *          *

UPPER 95% C.I.   -9.45E-01   4.04E+01   5.87E+01   4.34E-03   1.35E-03
FINAL VALUES     -1.07E+00   3.41E+01   5.05E+01   1.25E-03   1.14E-03
LOWER 95% C.I.   -1.20E+00   2.78E+01   4.22E+01   3.59E-04   9.67E-04

    REASONABLE
    UPPER LIMIT  -8.00E-01   8.00E+01   6.00E+01   1.20E-02   1.30E-02
    REASONABLE
    LOWER LIMIT  -1.40E+00   3.00E+01   2.00E+01   1.20E-04   1.30E-04

ESTIMATE ABOVE (1)
BELOW(-1)LIMITS      0           0          0          0          0
ENTIRE CONF. INT.
ABOVE(1)BELOW(-1)    0           0          0          0          0
```

```
NATIVE PARAMETER VALUES --- EXP10 OF LOG TRANSFORMED PARAMETERS

PARAMETER:          HK_1       VK_CB      SS_2       HK_2
* = LOG TRNS:        *          *          *          *

UPPER 95% C.I.      4.80E-04   2.79E-07   3.08E-04   5.48E-05
FINAL VALUES        4.26E-04   2.17E-07   6.03E-05   4.82E-05
LOWER 95% C.I.      3.78E-04   1.70E-07   1.18E-05   4.24E-05

      REASONABLE
      UPPER LIMIT   3.00E-03   1.00E-06   2.00E-03   4.00E-04
      REASONABLE
      LOWER LIMIT   3.00E-05   1.00E-08   2.00E-05   4.00E-06

ESTIMATE ABOVE (1)
BELOW(-1)LIMITS        0          0          0          0
ENTIRE CONF. INT.
ABOVE(1)BELOW(-1)      0          0          0          0


   LEAST-SQUARES OBJ FUNC (OBS. ONLY)----- =  23.908
   LEAST-SQUARES OBJ FUNC (W/PRIOR)------- =  23.908
   NUMBER OF INCLUDED OBSERVATIONS-------- =      35 OF       35
   NUMBER OF PRIOR ESTIMATES-------------- =       0
   NUMBER OF ESTIMATED PARAMETERS--------- =       9

   CALCULATED ERROR VARIANCE (CEV)-------- = 0.91953
   95% CONFIDENCE INTERVAL ON CEV--------- = 0.57032      1.7274
   STANDARD ERROR OF THE REGRESSION------- = 0.95892
   95% CONFIDENCE INTERVAL ON STD ERR----- = 0.75520      1.3143

   CORRELATION COEFFICIENT---------------- = 0.99999
        W/PRIOR-------------------------- = 0.99999
   ITERATIONS---------------------------- =       5

MAX LIKE OBJ FUNC (MLOF)-- = -35.194

MODEL EVALUATION MEASURES RELATIVE TO MLOF:
AIC STATISTIC------------- = -9.9935
BIC STATISTIC------------- = -3.1954
HQ STATISTIC-------------- = -12.361
KASHYAP STATISTIC--------- =  22.447


N*LOG(SIGMA**2)----------- = -13.340

MODEL EVALUATION MEASURES RELATIVE TO N*LOG(SIGMA**2):
AIC STATISTIC------------- = 15.827
BIC STATISTIC------------- = 22.214
HQ STATISTIC-------------- = 12.029
KASHYAP STATISTIC--------- = 46.018

LOG DETERMINANT OF FISHER INFORMATION MATRIX = 42.184
```

253

```
SUMMARY OF PARAMETER VALUES AND STATISTICS FOR
     ALL PARAMETER-ESTIMATION ITERATIONS
--------------------------------------------------------------------------------
--------------------------------------------------------------------------------


SELECTED STATISTICS FROM MODIFIED GAUSS-NEWTON ITERATIONS

        # PARAMs                  MAX.            MAX. CHANGE   DAMPING
  ITER. ESTIMATED   PARNAM        CALC. CHANGE    ALLOWED       PARAMETER
  ----- ---------   ------------  ---------------- ------------  ------------
    1       9       RCH_2         0.465120         2.00000       1.0000
    2       9       RCH_2         0.129055         2.00000       1.0000
    3       9       K_RB          0.834048E-01     2.00000       1.0000
    4       9       K_RB          0.424556E-01     2.00000       1.0000
    5       9       K_RB          0.814845E-02     2.00000       1.0000


--------------------------------------------------------------------------------


PARAMETER VALUES FOR EACH ITERATION
- INDICATES DAMPING RESULTED FROM PARAMETER CONSTRAINTS FOR THAT ITERATION.
* INDICATES SOME PARAMETERS WERE OMITTED DURING THAT ITERATION.  THE VALUES
  OF THOSE PARAMETERS ARE REPORTED EVEN THOUGH THEY WERE NOT UPDATED

    ITER   Q1&2          RCH_1         RCH_2         K_RB          SS_1
     0    -1.100         60.00         30.00         0.1200E-02    0.1300E-02
     1    -1.019         39.39         43.95         0.2047E-03    0.1242E-02
     2    -1.062         34.72         49.63         0.4910E-03    0.1154E-02
     3    -1.074         34.15         50.45         0.9270E-03    0.1141E-02
     4    -1.074         34.13         50.48         0.1247E-02    0.1141E-02
     5    -1.074         34.13         50.49         0.1317E-02    0.1140E-02
  FINAL   -1.074         34.13         50.48         0.1247E-02    0.1141E-02

    ITER   HK_1          VK_CB         SS_2          HK_2
     0    0.3000E-03    0.1000E-06    0.2000E-03    0.4000E-04
     1    0.4001E-03    0.1902E-06    0.8850E-04    0.4346E-04
     2    0.4237E-03    0.2174E-06    0.5328E-04    0.4737E-04
     3    0.4256E-03    0.2172E-06    0.6043E-04    0.4819E-04
     4    0.4258E-03    0.2175E-06    0.6026E-04    0.4820E-04
     5    0.4258E-03    0.2174E-06    0.6034E-04    0.4820E-04
  FINAL   0.4258E-03    0.2175E-06    0.6026E-04    0.4820E-04
--------------------------------------------------------------------------------


SUMS OF SQUARED WEIGHTED RESIDUALS FOR EACH ITERATION

          SUMS OF SQUARED WEIGHTED RESIDUALS
          ITER.   OBSERVATIONS   PRIOR INFO.      TOTAL        # INCLUDED OBS
             0    0.13406E+06    0.0000        0.13406E+06     35 OF      35
             1    672.17         0.0000        672.17          35 OF      35
             2    37.050         0.0000        37.050          35 OF      35
             3    26.322         0.0000        26.322          35 OF      35
             4    23.908         0.0000        23.908          35 OF      35
             5    23.908         0.0000        23.908          35 OF      35
          FINAL   23.908         0.0000        23.908          35 OF      35

********************************************************************************
********************************************************************************
  Parameter Estimation CONVERGED: % change of PARAMETER VALUES less than TolPar
********************************************************************************
```

****

```
Run end date and time (yyyy/mm/dd hh:mm:ss): 2006/01/09 22:26:06
Elapsed run time:  8.672 Seconds
```

Selected Data-Exchange Files

The header line and the next five lines of data-exchange files with file extensions _os, _w, and
_ws are shown. Some header lines has been reformatted to fit on one line in the table.

ex1._os				
"SIMULATED EQUIVALENT"	"OBSERVED or PRIOR VALUE"	"PLOT SYMBOL"	"OBSERVATION…"	
100.1854	101.8040	1	h1.0	
-0.8392334E-04	-0.2900000E-01	1	h1.1	
-0.8770752E-01	-0.1290000	1	h1.12	
126.9964	128.1170	1	h2.0	
-0.3368378E-01	-0.4100000E-01	1	h2.1	
ex1._w				
"WEIGHTED RESIDUAL"	"PLOT SYMBOL"	"OBSERVATION or PRIOR NAME"		
1.616581	1	h1.0		
-0.4089351	1	h1.1		
-0.5839639	1	h1.12		
1.119202	1	h2.0		
-0.1034670	1	h2.1		
ex1._ws				
"SIMULATED EQUIVALENT"	"WEIGHTED RESIDUAL"	"PLOT SYMBOL"	"OBSERVATION…"	
100.1854	1.616581	1	h1.0	
-0.8392334E-04	-0.4089351	1	h1.1	
-0.8770752E-01	-0.5839639	1	h1.12	
126.9964	1.119202	1	h2.0	
-0.3368378E-01	-0.1034670	1	h2.1	

Many of the data-exchange files are designed for plotting on x-y graphs. Figure C-2 shows the
Cook's D from the ex1._rc data-exchange file produced by the code RESIDUAL_ANALYSIS.
The results show that for the transient problem the head at point 8 of figure C-1B at 12 days into
the pumping test is most important to the estimated parameter values. In contrast to the
calibration of the steady-state system without pumpage presented by Hill and Tiedeman (in
press, table 7.2), the streamflow gain flow measurements are not very important. This is expected
because the pumpage is such a large stress on the system.

Figure C-3 shows the contents of the ex1._rdadv file plotted using the USGS open-source,
public-domain program GWCHART (Winston, 2000). The weighted residuals are all within their
associated theoretical intervals, so these results do not contradict the hypothesis that the model fit
is random and consistent with normally distributed true errors.

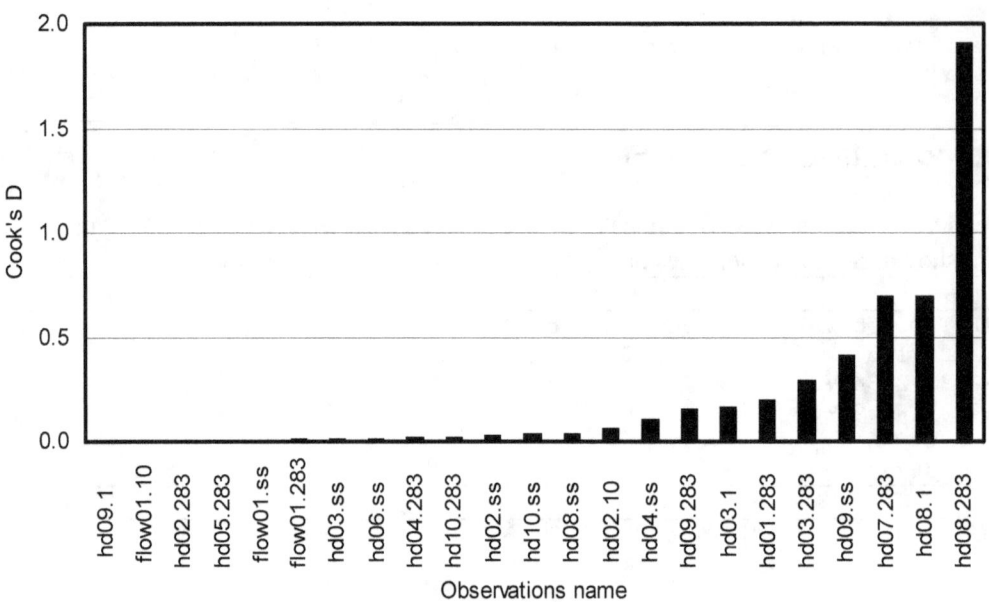

Figure C- 2. Cook's D calculated for the observations of the transient calibration. The largest 23 values are shown. Observations names that begin with 'hd' and end with '.ss' are for heads at steady state. Other observations names that begin with 'hd' are drawdowns. For heads and drawdowns, the well number from figure C-1B follows the 'hd'. Observation names that begin with 'flow' are streamflow gain measurements. The number at the end of many names is the number of days after pumping started.

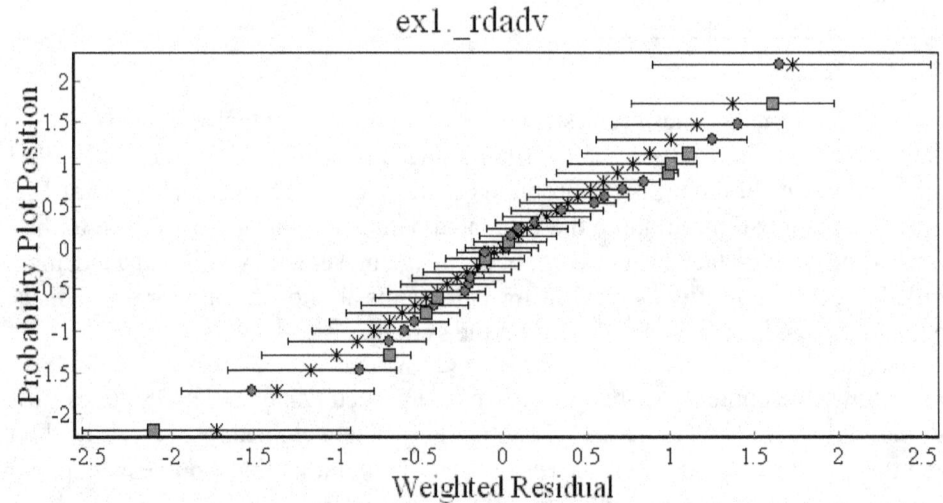

Figure C- 3. Plot of the contents of the ex1._rdadv file from the ex1a directory (see AppendixD) created using GWCHART (Winston, 2000). Heads are represented by squares, drawdowns are represented by circles, and flows are represented by triangles. The means of the theoretical weighted residuals are represented by an asterisk.

Particle-Path Predictions and Measures of Uncertainty

The movement of a particle from the X in the northeast corner of the area (figure C-1A) is predicted under steady-state conditions with pumpage. The ADV2 Package (Anderman and Hill, 2001) of MODFLOW-2000 is used. This package simulates particle motion as distance traveled in the three coordinate directions. Particle positions are calculated for four times: 10, 50, 100, and 175 years.

Porosity is not estimated because the observations of head, drawdown, and flow do not depend on porosity. However, the advective-transport prediction does depend on porosity. The uncertainty in porosity of the system can be included in measures of uncertainty of the particle position by defining one or more porosity parameters. Here, a porosity parameter is defined and named POR_1&2. It defines the porosity in aquifers 1 and 2 of figure C-1A. The porosity value is assigned to be 0.33. Prior information is applied with a standard deviation of 0.03. Thus, assuming a normal probability distribution, a 95-percent linear confidence intervals on the porosity is bound by the limits of 0.27 and 0.39, which is reasonable for the unconsolidated deposits involved. The porosity of the confining layer is set to 0.10 and is not defined using a parameter. The uncertainty in this value is not expected to be important, because the distance through the confining layer is small and does not affect the primary question posed in this problem, which is whether the particle goes to the well. Thus exclusion of this parameter is not expected to adversely affect the analysis.

To demonstrate the affect of including the uncertainty of the porosity in the analysis of prediction uncertainty, figure C-4 shows the width of 95-percent linear confidence intervals that exclude and include a porosity parameter. The linear confidence intervals are calculated using the code LINEAR_UNCERTAINTY and are listed in the _linp data-exchange file. For some of the predictions, confidence interval widths increase substantially when the porosity parameter is included. The linear intervals without the porosity parameter are produced in subdirectory ex1a; the intervals with the porosity parameter are produced in subdirectory ex1b (see Appendix D).

The predicted particle paths and linear and nonlinear individual confidence intervals for motion in the two planar coordinate directions are shown in figure C-5. The linear confidence intervals are calculated using LINEAR_UNCERTAINTY. The nonlinear confidence intervals are calculated using the UCODE_2005 nonlinear-uncertainty mode. The porosity parameter is included and the files needed to produce these results are provided in the ex1b subdirectory (see Appendix D).

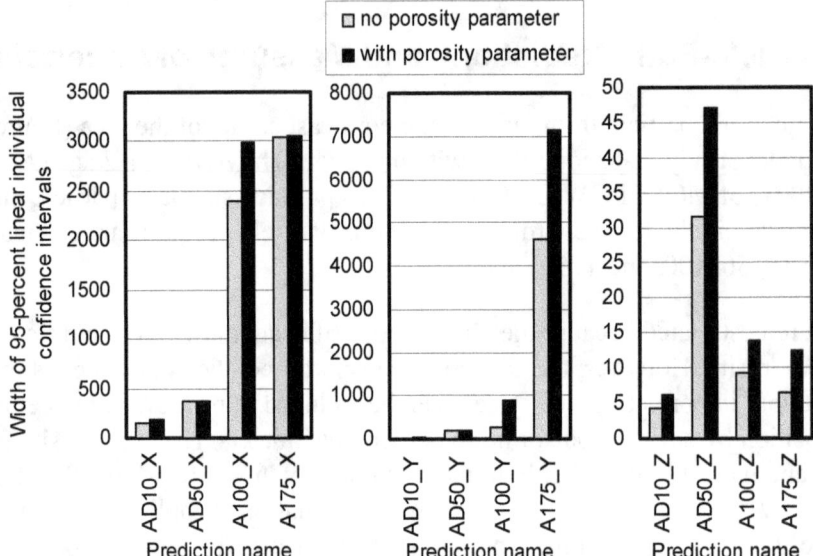

Figure C- 4. Width of 95-percent linear confidence intervals for the predicted advective transport of a particle in the three coordinate directions at 10, 50, 100, and 175 years without and with a porosity parameter. Results without and with a porosity parameter are produced using files in directory ex1a and ex1b, respectively (see Appendix D). In each prediction name, the number is the number of years of transport and the last letter indicates transport in the x, y, or z direction.

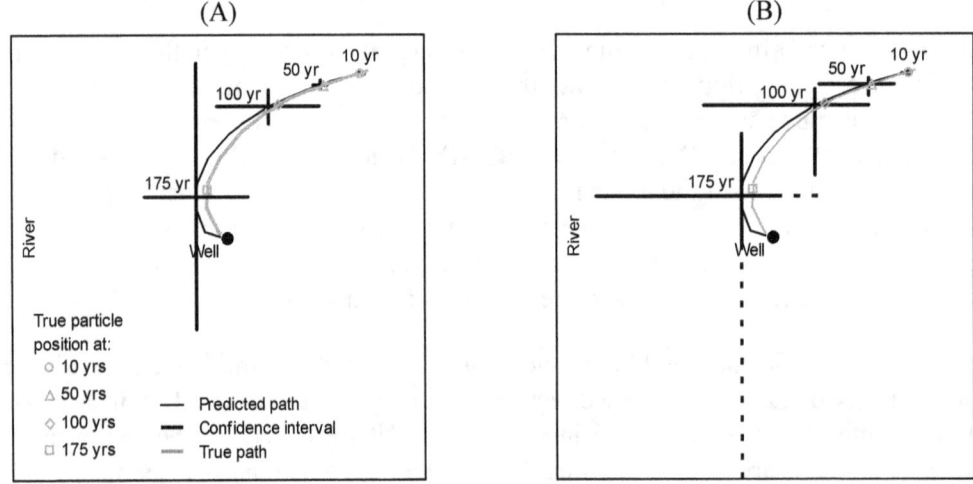

Figure C- 5: Plan view showing predicted and true advective-transport paths and particle locations at travel times 10, 50, 100, and 175 years. Simultaneous, 95-percent confidence intervals are shown calculated using (A) the linear methods of LINEAR_UNCERTAINTY and (B) the nonlinear methods of the UCODE_2005 nonlinear-uncertainty mode. The x direction is horizontal on the page; the y direction is vertical. At 175 years the dashed lines in (B) are affected by the projection simulated by the ADV2 Package when a particle exits the simulated system. Here, the particle exits the well. The realistic interval limit is the well location, where the solid line ends. (from Hill and Tiedeman, in press, Chapter 9).

Appendix C: Example Simulation
-- Particle Path Predictions and Measures of Uncertainty --

Table C-3 shows that total model nonlinearity, as measured using Beale's measure and the total model nonlinearity statistic, is large. However, intrinsic model linearity is much smaller, indicating that some statistics, such as the Cook's D measure shown in figure C-2, are not adversely affected by nonlinearity.

Table C- 3. Nonlinearity measures for the transient problem with 10 defined parameters, including porosity.

Name of Linearity Measure	Value	Critical Value	Output file[1]
Modified Beale's measure	93	[2]Nonlinear if > 0.45	ex1.#modlin
Intrinsic nonlinearity measure	0.62	[2]Linear if <<24	ex1.#resanadv
Total model nonlinearity	45	[3]Nonlinear if >1.0	ex1.#modlinadv
Intrinsic model nonlinearity	0.13	[3]0.09 to 1.0 nonlinear	ex1.#modlinadv

[1] These are the file extensions used by UCODE_2005. The batch files in the ex1b subdirectory (see Appendix D) change the filenames to add run sequence number after the #.
[2] Critical value is problem dependent. It is printed in the output file with an interpretive statement.
[3] Critical values are presented in table 38.

The combined intrinsic nonlinearity is printed at the end of the file with file extension #modlinadv.

Objective-Function Surface for the Steady-State Problem with Two Parameters

Figure C-5 shows the objective-function surface produced for the steady-state version of the problem shown in figure C-1 without pumpage using the starting parameter values listed in tables C-1 and C-2. Such plots can be created using the data-exchange file with filename extension _sos produced by the investigate-objective-function mode of UCODE_2005 (table 3).The plots are often used to investigate difficulties with regression. For this plot, the six parameters that apply of the nine listed in tables C-1 and C-2 are lumped into two parameters. Parameter RCH_MULT multiplies the values of the RCH_1 and RCH_2 parameters. Parameter K_MULT multiplies the values of the K_RB, HK_1, VK_CB, and HK_2 parameters.

The recharge and hydraulic-conductivity parameters are grouped because flows and hydraulic conductivities are often extremely correlated in ground-water problems. For the steady-state problem considered here, the two lumped parameters are completely correlated when only hydraulic-head observations are considered, as discussed by Hill and Tiedeman (in press, exercise 4.1). As parameters are lumped, parameter correlation can become easier to identify (Hill and Tiedeman, in press, section 4.4.2).

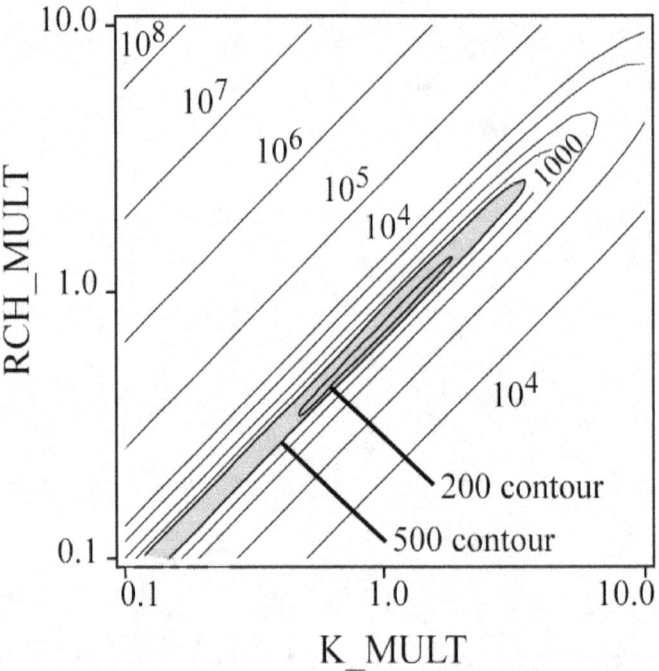

Figure C- 6. Objective-function surface for the steady-state problem with no pumpage when the six parameters that apply are lumped into two parameters. The data file used to produce this figure is the data-exchange file with extension _sos produced using the UCODE_2005 investigate-objective-function mode (table 3). The UCODE_2005 files are distributed in subdirectory ex1-ss-sos (Appendix D). (modified from Hill and Tiedeman, in press, figure 4-4.).

References

Anderman, E.R. and Hill, M.C., 2001, MODFLOW-2000, the U.S. Geological Survey modular ground-water model -- Documentation of the ADVective-Transport observations (ADV2) Package: U.S. Geological Survey Open-File Report 01-54, 69p. http://water.usgs.gov/nrp/gwsoftware/MODFLOW-2000/MODFLOW-2000.html

Harbaugh, A.W., Banta, E.R., Hill, M.C., and McDonald, M.G., 2000, MODFLOW-2000, the U.S. Geological Survey modular ground-water model, User's guide to the Modularization concepts and the Ground-Water Flow Process: U.S. Geological Survey Open-File Report 00-92, 121 p. http://water.usgs.gov/nrp/gwsoftware/MODFLOW-2000/MODFLOW-2000.html

Hill, M.C., 1998, Methods and guidelines for effective model calibration: U.S. Geological Survey, Water-Resources Investigations Report 98-4005, 90p.

Hill, M.C., Banta, E.R., Harbaugh, A.W., and Anderman, E.R., 2000, MODFLOW-2000, the U.S. Geological Survey modular ground-water model, User's guide to the Observation, Sensitivity, and Parameter-Estimation Process and three post-processing programs: U.S. Geological Survey Open-File Report 00-184, 209 p. http://water.usgs.gov/nrp/gwsoftware/MODFLOW-2000/MODFLOW-2000.html

Hill. M.C. and Tiedeman, C.R., in press, Effective model calibration, with analysis of data, sensitivities, predictions, and uncertainty: Wiley and Sons, New York, New York.

Appendix C: Example Simulation
-- References --

Appendix D. PROGRAM DISTRIBUTION AND INSTALLATION

Distributed Files and Directories

UCODE_2005 and the post-processors can be downloaded from the web site listed in the preface. The operating system is listed for each compiled downloadable executable file. When uncompressed, a directory is created with six subdirectories. The directory is named wrdapp\ucode_2005. The subdirectories are listed in table D-1. For windows operating systems, the batch files distributed in the test-win\ex1a subdirectory are listed in table D-2 and the batch files distributed in the test-win\ex1b subdirectory are listed in table D-3.

Compiling and Linking

If changes to the source codes are needed, or if the codes are used with an operating system other than those for which executable files are distributed, the codes need to be compiled. The modules needed to compile each of the distributed codes are listed in a readme file located in the subdirectory for each code. These subdirectories are located within the src subdirectory.

The distributed source code is compatible with standard Fortran 90 and Fortran 95 except for the following

(1) The call to the SYSTEM subroutine, which is used to initiate execution of an operating-system command. This call is in subroutine UTL_SYSTEM.

(2) The call to the GETCL subroutine, which provides access to the command line used to invoke a program. This call is in subroutine UTL_GETARG

These subroutines are non-standard and compiler-dependent. Both subroutines are in the utilities module (UTL.F90) of the JUPITER API. It is expected that any changes needed to accommodate compilers that have different subroutines or different syntax for these capabilities would be restricted to these subroutines.

The object files created during compilation need to be linked to create an executable program. The linker program commonly is invoked as part of the compilation procedure.

263

Appendix D. Program Distribution and Installation

Table D- 1: Contents of the subdirectories distributed with MODFLOW-2000.
[Name files, files used by MODFLOW-2000 to define program performance and input files.]

Subdirectory	Contents		
bin	Executable files of UCODE-2005, JRUNNER, and the six codes RESIDUAL_ANALYSIS, LINEAR_UNCERTAINTY, MODEL_LINEARITY, RESIDUAL_ANALYSIS_ADV, MODEL_LINEARITY_ADV, and CORFAC_PLUS. Also, an executable for MODFLOW-2000 (MF2K). Executables are platform dependent; the platform required is stated on the distribution site. These files can be executed by typing the file name at the operating-system command prompt or other methods supported by the operating systems. For example, batch files can be used on a Windows operating system.		
doc	This documentation file, in PDF format.		
src	Fortran source files organized into nine subdirectories. All source files are named with the extension "f90". Except for API-MODULES, each subdirectory contains a file called readme.txt that lists the modules needed to compile the code to which the directory is dedicated. The subdirectories are: **API-MODULES**: JUPITER API modules used in the codes listed below. **UCODE_2005**: Files unique to UCODE_2005. **RUNNER**: Files unique to the computer code RUNNER, which needs to be executed on machines to be used for parallel computations. Six directories each named for one of the other six codes documented in this report. Each of these directories contains the files unique to those codes.		
test-data-os[1]	Subdirectories contain a Name file, input files, and a run file (batch files when *os*=win) for a process-model forward run. Some also contain files to calculate sensitivity-equation sensitivities. After the programs are run in the test-*os* directory (see below), these subdirectories also include process-model output files.		
	Subdirectory	Contents	
	data-transient	Transient response to pumpage. Nine parameters.[2] Transient observations.[3]	
	data-adv-preds-ex1a	Steady-state with pumpage. Nine parameters.[2] Predictions.[4]	
	data-adv-preds-ex1b	Steady-state with pumpage. Ten parameters.[5] Predictions.[4]	
	data-ss	Steady state without pumpage. Steady-state observations.[6]	
	data-used-by-all	Files referred to by the Name files in the above directories.	
test-*os*[1]	Subdirectories contain run files, UCODE_2005 main input files, and other files for the test cases described in Appendix C. Any additional test cases are described in file cases.txt in this directory.		
	Subdirectory	Contents	
	ex1a ex1a-files	Batch files, UCODE_2005 main input files, and auxiliary files for results shown in table C-1 and figures C-2, C-3, and C-4.	
	ex1b ex1b-files	Batch files, UCODE_2005 main input files, and auxiliary files for results shown in figure C-4 and C-5.	
	ex1-ss-sos	Files for the shown in figure C-6.	
	ex1-true	Files for the results shown in table C-2.	

[1] The content of this directory is operating-system dependent. The directory name is formed by substituting the operating-system name for "*os*". For example, test-win includes files for a Windows operating system.

[2] The nine defined parameters are listed in tables C-1 and C-2.

[3] Observations are heads and drawdown at the 10 locations shown in figure C-1B and the streamflow gain at steady-state with no pumping and two times after pumping begins.

[4] Predictions are the x-, y-, and z- distances traveled at four times by a particle released from the location of the proposed landfill of figure C-1A.

[5] A porosity parameter is defined in addition to the nine parameters listed in tables C-1 and C-2. The Porosity parameter is named Por_1&2 and applies to aquifers 1 and 2 of figure C-1A. The value equals 0.33. Prior information is applied with a standard deviation of 0.3. The porosity of the confining layer equals 0.10 and is not defined using a parameter.

[6] Observations are heads at the 10 locations shown in figure C-1B and the streamflow gain.

Portability

UCODE_2005 and the post-processors were written in standard Fortran 90/95. A modular style is used to enhance accuracy, to simplify maintenance, and to encourage innovation.

The aspects of the code mentioned above that are compiler dependent are also platform dependent.

Memory Requirements

As distributed, the source files and executable file dynamically allocate memory. Thus, the program automatically adapts to whatever memory is required and no user intervention is required. Slow execution times can result if the memory required exceeds the physical memory available on the computer being used.

Table D- 2. The batch files distributed in test-win subdirectory ex1a, in which nine
        parameters are defined.

[Running these batch files produces the results shown in table C-1. Main output files are
renamed in the batch files. For example, for the UCODE_2005 parameter-estimation
mode run produced by batch file 03-parest+resan+resanadv.bat, the ex1.#uout file is
renamed as ex1.#03uout-parest.]

Batch file name	Purpose
00-test-command.bat	Test command used in the Model_Command_Lines input block.
01-ucode-forward.bat	Execute UCODE_2005 forward mode using starting parameter values.
02-ucode-sen-analysis.bat[1]	Execute UCODE_2005 sensitivity-analysis mode.
03-parest+resan+ resanadv.bat	Execute UCODE_2005 parameter-estimation mode using sensitivity-equation sensitivities produced by MODFLOW-2000, and execute RESIDUAL_ANALYSIS, RESIDUAL_ANALYSIS_ADV.
04-ucode-modlin+ modlin.bat	Execute UCODE_2005 test-model-linearity mode and MODEL_LINEARITY.
05-ucode-pred.bat	Execute UCODE_2005prediction-mode. Predictions are advective transport simulated using ADV2. Sensitivities are obtained from UCODE-2005 using perturbation.
06-linunc.bat	Execute LINEAR_UNCERTAINTY.
07-03-to-06.bat	Perform parameter-estimation, calculate predictions, and execute the other four programs listed above in this table.
08-clean_process- models.bat	Delete all files in the subdirectories of the test-data-win directory created by the above runs.
09-clean_all.bat	Delete all files created with any of the batch files. This includes files created in the test-data-win subdirectories as well as the ex1a subdirectory.

[1] The main output file of this run ends with an error message saying that the _b1 data-exchange
file is not written. The starting parameter values used for this run result in the parameter values
for Beale's measure that are normally written to the _b1 file to be mathematically invalid. The
file is produced correctly after regression is performed.

Table D- 3. Batch files distributed in test-win subdirectory ex1b, in which 10 parameters are defined, including a porosity parameter important to predictions.
[Running these batch files produces the results shown in figure C-5B, table C-3, and figure C-6. All sensitivities are calculated by UCODE_2005 using perturbation methods. Main output files are renamed in the batch files. For example, for the UCODE_2005 sensitivity-analysis mode run produced by batch file 11-ucode-sen-analysis.bat, the ex1.#uout file is renamed as ex1.#03uout-parest.]

Batch file name	Purpose
11-ucode-sen-analysis.bat	Execute UCODE_2005 sensitivity-analysis mode.
12-ucode-sen-analysis-init.bat	Execute UCODE_2005 sensitivity-analysis mode with keyword CreateInitFiles=yes in the UCODE_Control_Data input block.
13-resanadv.bat	Execute RESIDUAL_ANALYSIS_ADV.
14-ucode-modlin+modlin.bat	Execute UCODE_2005 test-model-linearity mode and MODEL_LINEARITY.
15-ucode-pred.bat	Execute UCODE_2005 prediction-mode. Predictions are advective transport simulated using ADV2.
16-linunc.bat	Execute LINEAR_UNCERTAINTY to obtain linear intervals that include uncertainty with which the porosity is known.
17-corfac.bat	Execute CORFAC with ConfidenceOrPrediction=Confidence. Thus, nonlinear confidence intervals are calculated in step 19.
18-ucode-modlinadv+modinadv.bat	Execute UCODE_2005 advanced-test-model-linearity mode.
19-ucode-nonlinear-intervals.bat	Delete all files in the subdirectories of the test-data-os directory used in the above runs.
20-all.bat	Perform all analyses listed above in this table.
21-clean_process-models.bat	Delete all files in the subdirectories of the test-data-win directory crated by the above runs.
22-clean_all.bat	Delete all files created with any of the batch files. This includes files created in the test-data-win subdirectories as well as the ex1a subdirectory.

Appendix D. Program Distribution and Installation

Appendix E. Comparison with UCODE as Documented by Poeter and Hill (1998)

The aspects of the computer codes documented in this work that are similar, different, and new relative to UCODE as documented by Poeter and Hill (1998) are listed in table E-1. The table is largely self-explanatory, but the input file design for UCODE_2005 deserves some additional explanation here. All UCODE_2005 features are described thoroughly in the main part this report.

The input file design for UCODE_2005 differs substantially from the input file design for UCODE. This arises primarily because UCODE_2005 is written entirely in Fortran90/95, while UCODE was written using a combination of Perl and Fortran77. As a result of Dr. John Doherty's participation with the JUPITER API used to construct UCODE_2005, some aspects of the UCODE_2005 input files are similar to PEST input files (Doherty, 2004). For example, the methods available in UCODE_2005 to extract values from process-model output files is identical to that used in PEST except that the UCODE_2005 Standard File option for reading has been added. Also, the equation protocols used to define derived observations and parameters and to define prior information are identical to those used in PEST, and template files are identical except for the three letters used on the first line of each template file.

Appendix E. Comparison with UCODE as Documented by Poeter and Hill (1998)

Table E- 1. UCODE_2005 compared to UCODE as described in Poeter and Hill (1998).
[UCODE 2005 input blocks are described in Chapters 5 to 13 of this report.]

Capabilities that are the same or very similar
S1. The modified Gauss-Newton method except as described under New Capabilities.
S2. Process model(s) need to be run with batch files.
S3. Interaction with process model(s) is through text only input and output files.
S4. Flexible use of parameters to produce model input files and use of extracted values to produce simulated equivalents to observations. The format of the equations has changed.
S5. Produce data sets for visualizing objective functions.

Capabilities that are different
D1. **Input files.** UCODE input files are replaced by input blocks in the main UCODE_2005 input file and other files as follows:
UCODE Universal file ➔ UCODE_Control_Data, Reg_GN_Controls, Model_Command_Lines, Observation input blocks.
UCODE Prepare file ➔ Parameter, Model_Input_Files input blocks.
Template input files ➔ Template input files that are similar.
UCODE Function file ➔ Equation capabilities of the Parameter input blocks.
UCODE Extract file ➔ Model_Output_Files input block, Instruction input files.
D2. **Extraction.** Except for new Standard File option, extraction is as in PEST (Doherty, 2004).
D3. **Data-exchange files.** Many UCODE output files with filename suffixes of the form _xxx are now data-exchange files, which are different in that (a) all have a header line with column labels, (b) some have been eliminated, and (c) there are many new files (see Chapters 14 and 16). The _ws file now contains simulated values instead of weighted simulated values.
D4. **Modes instead of Phases.** UCODE phases in parentheses and UCODE_2005 modes: (1) forward mode. (11) investigate-objective-function mode. (2, 22) sensitivity-analysis mode. (3) parameter-estimation mode with RESIDUAL_ANALYSIS. (44, 45) prediction mode with LINEAR_UNCERTAINTY. (33) test-model-linearity mode with MODEL_LINEARITY.

Capabilities that are new
N1. **Sensitivities.** Can use sensitivities produced by process model(s). For example, MODFLOW-2000 sensitivity-equation sensitivities (Hill and others, 2000) can be used.
N2. **Non-detects.** For observations (usually concentrations) below a detection limit.
N3. **No solution.** Use for process models that may not be able to produce simulated equivalents and that write a distinct value to output files. For example, cells may go dry.
N4. **Weighting with simulated values.** Weights can be calculated using coefficients of variation and simulated values. For example, use this to weight concentration observations.
N5. **Full weight matrix.** For groups of observations or prior information.
N6. **Faster, more robust parameter estimation.** In tests, the trust-region modification to Gauss-Newton reduced the number of parameter-estimation iterations to about half and was successful in more situations (Mehl and Hill, 2003).
N7. **Parallel processing.** Send parameter-loop simulations (Figure 1) to separate processors.
N8. **Standard Files.** Easily extract values from columns of numbers.
N9. **Unique criteria for each parameter.** The criteria govern (1) the maximum fractional parameter-value change in one parameter-estimation iteration and (2) the fractional parameter-value change allowed when parameter-estimation is said to converge. For the two criteria, smaller and larger values, respectively, may be useful for insensitive parameters.
N10. **Dynamic omission of insensitive parameters.** Remove insensitive parameters from regression calculations. Reevaluates each parameter-estimation iteration.
N11. **Constrained parameter values.** Use if parameter values result in process-model failure. Avoid use to ensure reasonable parameter estimates, which can be used to detect model error.
N12. **Nonlinear confidence intervals** and other capabilities described in Chapter 17.

Appendix F. ABBREVIATED INPUT INSTRUCTIONS FOR UCODE_2005

The abbreviated input instructions listed here are intended for quick reference by experienced users. Complete input instructions are presented in Chapters 6 through 13 and 17. The input blocks are presented here in the order they need to appear in the UCODE_2005 main input file. The chapter where the input block is described is noted.

Options Input Block (optional) Chapter 6

Verbose — Controls printing to the UCODE_2005 main output file.

Derivatives_Interface - Filename or path of file defining how to read derivatives.

PathToMergedFile - Filename or path of merged file. If the file exists, it is replaced.

Merge_Files Input Block (Optional) Chapter 6

PathToFile - Path of a file.

SkipLines - Lines to skip at the top before file is appended. Default=0.

UCODE_Control_Data Input Block (optional) Chapter 6

ModelName - Identifies the model. Up to 12 characters. Default=generic.

ModelLengthUnits - Defines the LENGTH units. Up to 12 characters. Default=NA.

ModelMassUnits - Defines the MASS units. Up to 12 characters. Default=NA.

ModelTimeUnits - Defines the TIME units. Up to 12 characters. Default=NA.

For the following seven variables, see Table 3 for modes produced by "yes".

Sensitivities - yes: calculate sensitivities. Default is no.

Optimize - yes: estimate parameters. Default is no.

Linearity - yes: do the calculations and produce file fn._b2. Default=no.

Prediction - yes: determine predictions and their sensitivities. Default=no.

LinearityAdv - conf, pred, or no. Default=no.

NonlinearIntervals - yes: calculate nonlinear confidence intervals. Default=no.

SOSsurface - yes or file: calculate objective-function values. Default=no.

SOSfile - file used when SOSsurface=file.

StdErrOne - yes: calculate statistics with s^2 replaced by 1.0. Default=no.

EigenValues - yes: calculate eigenvalues and eigenvectors. Default=yes.

Three keywords control printing of tables of observations, simulated values, and residuals to the main output file.

StartRes   - For the starting parameter values. Default=yes.

IntermedRes   - For parameter-estimation iterations. Default=no.

FinalRes   - For the final parameter values. Default=yes.

Three keywords control printing of sensitivity tables to the main output file.

StartSens   - For the starting parameter values. Default=dss.

IntermedSens   - For parameter-estimation iterations. Default=none.

FinalSens   - For the final parameter values. Default=dss.

DataExchange   - yes: generate the data-exchange files. Default=yes.

CreateInitFiles   - yes: generate only the _init data-exchange files. Default=no.

Reg_GN_Controls Input Block (optional) Chapter 6

Three keywords control when parameter-estimation iterations stop.

TolPar   - Tolerance based on parameter values. Default=0.01.

TolSOSC   - Tolerance based on model fit. Default=0.0.

MaxIter   - Maximum number of parameter-estimation iterations. Default=5.

Two keywords restrict how much parameter values can change in one parameter-estimation iteration.

MaxChange   - Maximum fractional amount parameter values are allowed to change between parameter-estimation iterations. Default=2.0.

MaxChangeRealm   - Indicates whether MaxChange applies in native or regression space. Default=Native.

Three keywords are used to calculate the Marquardt parameter

MqrtDirection   - Angle (in degrees) between down-gradient direction on the sum-of-squared-residuals surface and the parameter update vector. Default=85.4°.

MqrtFactor   - See equation 8 for the Marquardt parameter. Default=1.5.

MqrtIncrement   - See equation 8 for the Marquardt parameter. Default=0.001.

Three keywords control quasi-Newton updating

QuasiNewton   - yes: use quasi-Newton updating as indicated by the criteria below. Default=no.

If either of the following two criteria is met for a parameter-estimation iteration, Quasi-Newton updating is used for that and all subsequent iterations.

QNiter — Number of iterations executed before including the Quasi-Newton enhancement. Default=5.

QNsosr — Fractional change in the sum-of-squared weighted residuals over two parameter iterations below which Quasi-Newton matrix enhancement is employed. Default=0.01.

OmitDefault — The number of values to read from user-created file fn.omit. Default=0.

Stats_On_Nonconverge – yes: calculate final sensitivities and calculate and print statistics when parameter estimation does not converge in the maximum number of iterations. Default=yes.

Three keywords control dynamic omission of insensitive parameters from regression.

OmitInsensitive – yes: omit parameter j from the regression if its composite scaled sensitivity (CSS_j) satisfies $CSS_j <$ (MinimumSensRatio $\times$ CSSmax). Default=no.

MinimumSensRatio – Used as described for OmitInsensitive. Default=0.005.

ReincludeSensRatio – If ReincludeSensRatio>0.0 and $CSS_j >$ (ReincludeSensRatio $\times$ CSSmax), reinclude parameter j. Default=0.02.

TolParWtOS – TolParWtOS $\times$ TolPar equals the parameter-change threshold below which simulated values are used to calculate weights on observations with WtOSConstant>0. Default=10.

Four keywords control the trust-region modification of Gauss-Newton regression.

TrustRegion - Dogleg: use the double-dogleg modification. Default=no.

MaxStep - Maximum allowable step size used in the trust-region method. The default is a function of the sensitivities and the parameter values, and is printed in the UCODE_2005 main output file.

ConsecMax - Maximum number of times that MaxStep is used consecutively before execution stops. Default=5.

Reg_GN_NonLinInt Input Block (optional) Chapter 17

ConfidenceOrPrediction – confidence or prediction interval. Default=confidence.

IndividualOrSimultaneous – individual or simultaneous interval. Default=individual.

WhichLimits - Lower, Upper or Both. Default=Both.

TolIntP - Tolerance based on parameter values. Default=0.001.

TolIntS - Tolerance based on model fit. Default=0.1$\times$TolIntP.

TolIntY   - Tolerance based on change in the value of the computed interval limit. Default=0.001.

CorrectionFactors   - yes: Use correction factors. Default=no.

The following keyword is used only if CorrectionFactors=yes.

AlternateStartValues – yes: Use starting parameter values from Parameter_Values input block. Default=no, which means use values in fn._paopt.

Model_Command_Lines Input Block (required) Chapter 6

Command   - Operating system command that executes the process model(s).

Purpose   - The type of process model run performed. Default=forward.

CommandID   - A name for the command.

Parameter_Groups Input Block (optional) Chapter 7

GroupName   - The name of the group (up to 12 characters; not case sensitive). Default=ParamDefault

Other keywords   - Any keyword from the Parameter_Data input block.

Parameter_Data Input Block (required) Chapter 7

ParamName   - Parameter name (up to 12 characters; not case sensitive)

GroupName   - Group name (up to 12 characters; not case sensitive). Default=ParamDefault.

StartValue   - Starting parameter value. Default=A huge real number.

LowerValue   - Smallest reasonable value for this parameter. Default = -(A huge real number).

UpperValue   - Largest reasonable value for this parameter. Default = +(A huge real number).

Constrain   - yes: constrain the parameter value. Default=no.

UpperConstraint   - Upper limit on the parameter value.

LowerConstraint   - Lower limit on the parameter value.

Adjustable   - yes: this parameter value can be changed for the purpose defined in the UCODE_Control_Data input block. Default=no.

PerturbAmt   - Fractional amount of parameter value to perturb to calculate sensitivities for this parameter. Default=0.01.

Transform   - yes: log-transform the parameter for the regression. Default=no.

TolPar — Replaces, for this parameter, the value of TolPar from the Reg_GN_Controls input block or the default of 0.01.

MaxChange — Maximum fractional parameter change allowed between parameter iterations. Default=2.0.

SenMethod — how sensitivities are obtained. -1=read as log transformed, 0=read as native, 1=forward perturbation, 2=central perturbation. Default=1.

ScalePval — A positive number used to scale sensitivities if the parameter value gets too small. Default=StartValue/100.

SOSIncrement — The number of values to be considered when SOSsurface=yes in the UCODE_Control_Data input block. Default=5.

NonLinearInterval – yes: calculate nonlinear intervals for this parameter when NonlinearIntervals=yes in the UCODE_CONTROL_DATA block. Default=no.

Parameter_Values Input Block (optional) Chapter 7

ParamName — The name of the parameter for which a value is specified.

StartValue — The specified parameter value.

Derived_Parameters Input Block: (optional) Chapter 7

DerParName — Name of derived parameter (up to 12 characters; not case sensitive).

DerParEqn — An equation without an "equal" sign (that is, just the right-hand side of the equation) by which the derived parameter is calculated, generally using defined parameters.

Observations (omit for prediction mode) Chapter 8

Observation_Groups Input Block (optional)

GroupName — Name for a group of observations (up to 12 characters; not case sensitive). Default=DefaultObs.

UseFlag — yes: use the simulated values in this group to compare against observed values in the regression. Default=yes.

PlotSymbol — An integer intended for use in post-processing programs to assign symbols for plotting. Default=1.

WtMultiplier	- Value used to multiply weights for members of a group when the weights are defined using Statistic and StatFlag keywords of the Observation_Data input block. Default=1.0.
CovMatrix	- Name of the error variance-covariance matrix.
Other keywords	- Any keyword from the Observation_Data input block.

Observation_Data Input Block (required)

ObsName	- Observation name (up to 20 characters; not case sensitive). Each observation name needs to start with a letter and to be unique.
ObsValue	- Observation value.
Statistic	- Statistic used to calculate the observation weight.
StatFlag	- Defines Statistic. Options: VAR, SD, CV, WT, SQRWT. No default.
GroupName	- Group name from the Observation_Groups input block. Default=DefaultObs.
Equation	- An equation without an "equal" sign (just the right had side of the equation) that defines how to calculate an equivalent simulated value from simulated equivalents of previously defined observations. Default= _.
NonDetect	- Detection limit for an observation. Default=0.
WtOSConstant	- The constant η in equation 3. Default=0.

Derived_Observations Input Block (optional)

The Derived_Observations input block is identical to the Observation_Data input block except in name. It is included in UCODE_2005 so the user can define derived observations in a separate block, which is convenient in some circumstances.

Predictions (Omit for all modes but prediction, advanced-test-model-linearity, and nonlinear-uncertainty) Chapter 8

Prediction_Groups Input Block (optional)

GroupName	- Name for a group of predictions (up to 12 characters; not case sensitive). Default=DefaultPreds.
UseFlag	- yes: report and analyze the predictions in this group. Default=yes.
PlotSymbol	- An integer intended for use in post-processing programs to assign symbols for plotting. Default=1.
Other keywords	- Any keyword from the Prediction_Data input block.

Prediction_Data Input Block (required for three modes)

PredName — Prediction name (up to 20 characters; not case sensitive).Each prediction name needs to start with a letter and to be unique.

RefValue — Reference value to which the prediction is compared.

MeasStatistic — A statistic used to calculate the variance of the measurement error.

MeasStatFlag — Defines MeasStatistic. Options: VAR, SD. No default.

GroupName — Group name from the Prediction_Groups input block.

Equation — An equation without an "equal" sign (just the right hand side) that defines how to calculate a derived prediction. Default= '_'.

Derived_Predictions Input Block (optional)

The Derived_Predictions input block is identical to the Prediction_Data input block except in name. It is included in UCODE_2005 so the user can define derived predictions in a separate block, which may be convenient in some circumstances.

Prior_Information_Groups Input Block (optional) Chapter 9

GroupName — Name for a group of prior information items (up to 12 letters, numbers, and _; not case sensitive). Default=DefaultPrior.

UseFlag — yes: include this group when estimating parameters. Default=yes.

PlotSymbol — An integer used in post-processing programs for the purpose of assigning symbols for plotting. Default=1.

WtMultiplier — Value that multiplies the weights for members of the group when the weighting is defined using Statistic and StatFlag keywords described for the Linear_Prior_Information input block. Default=1.0

CovMatrix — Name of the error variance-covariance matrix.

Other keywords — Any keyword from the Linear_Prior_Information input block.

Linear_Prior_Information Input Block (optional) Chapter 9

PriorName — Prior information equation name (up to 20 letters, numbers, and _; not case sensitive; start with a letter). Default=DefaultPrior.

Equation — An equation without an "equal" sign that defines the prior information in terms of parameter names as specified in the Parameter_Data or Derived_Parameters input blocks.

PriorInfoValue — Value of prior information.

Statistic — Value used to calculate the prior information weight.

StatFlag	- Defines Statistic. Options: VAR, SD, CV, WT, SQRWT. No default.
GroupName	- Name for a group of prior information items.

Matrix_Files Input Block (optional) Chapter 10

MatrixFile	- Name or path of the file from which one or more matrices are read. (Up to 2,000 characters; case sensitivity depends on the operating system).
NMatrices	- Number of matrices to read from MatrixFile. Default=1.

Complete Matrix

```
CompleteMatrix  [NAME]
NGMEM  NGMEM  [ControlRecord]
[Array Control Record]
VAL(1,1) . . . . . . . VAL(1,NGMEM)
   .      .       .      .
   .      .       .      .
   .      .       .      .
VAL(NGMEM,1) . . . . . . . VAL(NGMEM,NGMEM)
```

Compressed Matrix

```
CompressedMatrix  [NAME]
NNZ  NGMEM  NGMEM [ControlRecord]
[Array Control Record]
IPOS(1)    VAL(1)
IPOS(2)    VAL(2)
...
IPOS(NNZ)   VAL(NNZ)
```

Array Control Record Input Instructions

1. *INTERNAL*   CNSTNT FMTIN IPRN
2. *OPEN/CLOSE* FNAME CNSTNT FMTIN IPRN

Model_Input_Files Input Block (required) Chapter 11

ModInFile	- Name for a process-model input file (up to 2,000 characters). Names with spaces need to be enclosed in double quotes.
TemplateFile	- Name for the template file (Up to 2,000 characters. Case sensitivity depends on the operating system). Names with spaces need to be enclosed in double quotes.

Template Files Chapter 11

A template file is created from a model input file by first inserting a line at the top. The line contains "jtf" followed by one or more spaces and the substitution delimiter. Commonly used substitution delimiters are @ and !.

The substitution delimiter is used to define the space within which UCODE_2005 places a number. The substitution space is defined by a pair of substitution delimiters. All of the characters between and including the substitution delimiters are replaced. Characters between the delimiters need to include spaces and one ParamName or DerParName defined in the Parameter_Data or Derived_Parameter input block. The ParamName or DerParName can be placed anywhere between the delimiters.

Model_Output_Files Input Block (required) Chapter 11

ModOutFile   - Name of the process-model output file with values to be extracted. Up to 2,000 characters; case sensitivity depends on the operating system.

InstructionFile   - Name for the Instruction file that UCODE_2005 uses to extract values from ModOutFile. InstructionFile can be up to 2,000 characters; case sensitivity depends on the operating system.

Category   - Identifies the type of quantity for which values are extracted. **Obs:** observations. **Pred:** predictions

Instruction Files (required) Chapter 11

For a Standard Process-Model Output File

```
jif @
StandardFile Nskip ReadColumn Nread
[Names for each of the Nread values.
Place each name on a new line.]
```

Nskip, number of lines to skip at top of file.
ReadColumn, the column to be read.
Nread, the number of items to be read.

For a Non-Standard Process-Model Output File

The first line of an instruction file needs to begin with the three letters "jif", a single space, and a marker delimiter. Usually $, @, % or ~ are good choices for marker delimiter.

Except for 'dum', which can be used repeatedly, a different name needs to be used for each extracted value.

A compete list of instructions is presented in Table 9 of Chapter 11.

Parallel_Control Input Block (Optional) Chapter 12

Parallel   - yes: Activates parallel processing. Default=no.

Wait   - Time delay, in seconds, used in file management. Default=0.001.

VerboseRunner   - Flag that controls printing by the runner. Default=3.

AutoStopRunners   -yes: stop runners when UCODE_2005 stops. Default=yes.

OperatingSystem   - Operating system for dispatcher and runner. Default=Windows.

TimeoutFactor   - Factor for RUNTIME to identify overdue run. Default=3.0.

Parallel_Runners Input Block (Optional) Chapter 12

RunnerName   - Name of runner. Up to 20 characters.

RunnerDir   - Pathname to directory where the runner program runs.

RunTime   - Expected model runtime, in seconds. Default=10.

Equation Protocols (optional) Chapter 13

In UCODE_2005, equations can be defined in the Derived_Parameters, Observation_Data, Derived_Observations, Prediction_Data, and Derived_Predictions input blocks. See Chapter 13 for additional information.

Derivatives Interface Input File (optional) Chapter 13

A Derivatives Interface input file provides UCODE_2005 with information needed to obtain model-calculated sensitivities (derivatives of simulated values with respect to parameters) from a model-output file rather than determining them by perturbation. See Chapter 13.

fn.xyzt Input File (optional) Chapter 13

The first line of the xyzt input file is ignored. The rest of the file needs to be composed of lines containing five columns of data: Observation name, x, y, z, and time.

Appendix G. ABBREVIATED INPUT INSTRUCTIONS FOR OTHER CODES

RESIDUAL_ANALYSIS (optional) Chapter 15

Nsets   - Number of sets. Default=4.

Seed   - Used to generate random numbers. Default=104857.

Calc_RandomNumbers - Create _rd and _rg output files. Default=yes.

Calc_DFBetas   - Calculate DfBetas statistics; create _rb output file. Default=yes.

The following keywords control printing to the #rs output file. Default=no.

Print_Par_Var_Cov_Matrix - Parameter variance-covariance matrix.

Print_Sqrt_Wt   - Square-root of the weight matrix.

Print_Unscaled_Sens - Unscaled sensitivities.

Print_CooksD   - Cook's D statistics.

Print_RB   - DFBetas statistics.

Print_RD   - Sets of uncorrelated random numbers.

Print_RG   - Sets of correlated random numbers.

Print_Res_Var_Cov_Matrix - Residual variance-covariance matrix

Print_Res_Correlation_Matrix - Residual correlation matrix.

RESIDUAL_ANALYSIS_ADV (optional) Chapter 17

Options Input Block (optional)

Verbose   - Controls printing to the #resanadv output file. Default=3.

RESIDUAL_ANALYSIS_ADV_Control_Data Input Block (optional)

Nsets   - Number of sets. Default=4.

Seed   - Used to generate random numbers. Default=104857.

StdDev   - Alternate standard error. Default=0.0.

Read_ET   - Read Mean_True_Error input block. Default=no.

Read_Cov   - Read matrix from the Matrix_Files input block. Default=no.

The following keywords control printing to the #resanadv output file. Default=no.

Print_IRMatrix   - Parameter (I-R) matrix.

Print_SimWgtResiduals – All sets of generated weighted residuals.

Print_Par_Var_Cov_Matrix - Parameter variance-covariance matrix.

Print_Sqrt_Wt   - Square-root of the weight matrix.

Print_Unscaled_Sens - Unscaled sensitivities.

Mean_True_Error Input Block (optional)

MTEName   - Observation name. No default.

MTEValue   - ET value. Default=0.0.

Matrix_Files Input Block (optional)

MatrixFile   - Filename or path of file containing a matrix.

NMatrices   - Number of matrices. Here, NMatrices=1.

CORFAC_PLUS (required) Chapter 17

Options Input Block (optional)

Verbose — Controls printing to the #resanadv output file. Default=3.

Correction_Factor_Data Input Block (required)

ConfidenceOrPrediction - Calculate confidence or prediction intervals. No default.

RegressionUsedTrueCov - Observation weights reflect observation errors. Default=no.

Read_Cov - Read matrix from the Matrix_Files input block. Default=no.

Effective_Correaltion - Upper limit of the spatial correlation. Default=0.8.

Read_ObsPred_Cov - Read second moments between observations and predictions using the Matrix_Files input block. Default=no.

Prediction_List Input Block (this or the next is required)

PredName - Prediction name. No default.

MeasStatistic - Statistic for measurement error. Default=variance from _pv file.

MeasStatFlag - Defines statistic. Options=VAR, SD. Default=VAR.

Parameter_List Input Block (this or the last is required)

ParamName - Parameter name. No default.

MeasStatistic - Statistic for measurement error. Default=variance from _pv file.

MeasStatFlag - Defines statistic. Options=VAR, SD. Default=VAR.

Matrix_Files Input Block (optional)

MatrixFile - Filename or path of file containing a matrix.

NMatrices - Number of matrices. Here, NMatrices=1 or 2.

www.ingramcontent.com/pod-product-compliance
Lightning Source LLC
Chambersburg PA
CBHW081433170526
45166CB00008B/2199